Key Concepts

STEM

SCIENCE TECHNOLOGY ENGINEERING MATH

PHYSICS

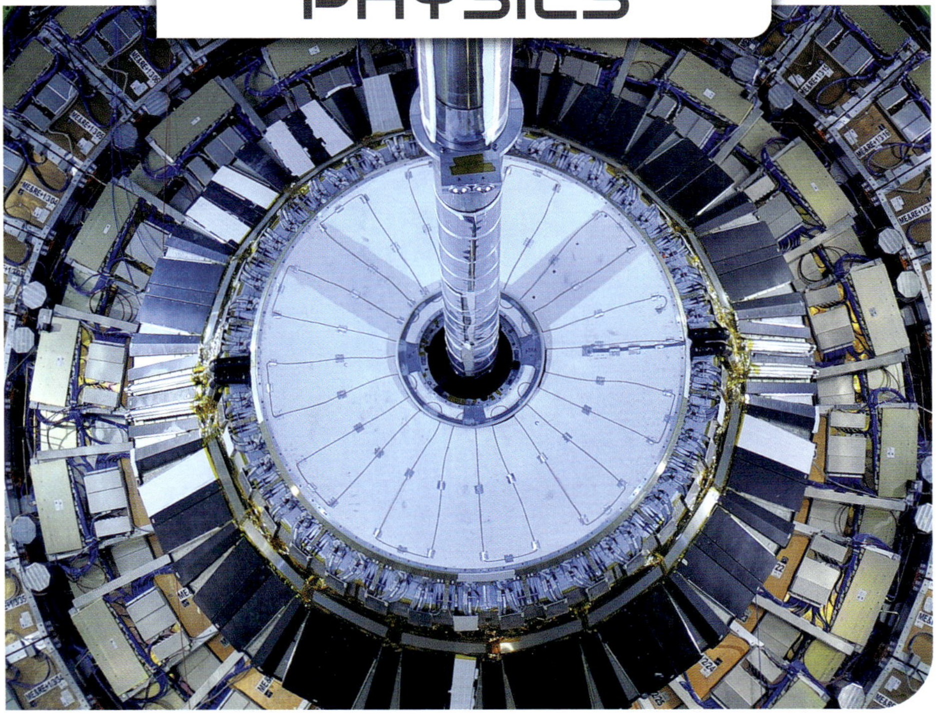

MAGNETISM

General Editor: Anne O'Dal

BROWN BEAR BOOKS

Published by Brown Bear Books Ltd
4877 N. Circulo Bujia, Tucson, AZ 85718
USA
and
Studio G14, Regent Studios, 1 Thane Villas,
London N7 7PH, UK

Cataloging-in-Publication data available upon request.

ISBN 978-1-83572-080-6 (ALB)
ISBN 978-1-83572-085-1 (paperback)
ISBN 978-1-83572-090-5 (ebook)

Children's Publisher: Anne O'Daly
Design Manager: Keith Davis
Picture Manager: Sophie Mortimer
Consultant: Chris Woodford

Manufactured in the United States of America
CPSIA compliance information: Batch#AG/5665

Picture Credits
Front Cover: Shutterstock: muratart: **Interior: iStock:** lagul 4, Hilda Weges 46–47; **Library of Congress:** 18b; media.daimler.com: 60–61; **Photos.com:** 32; **Public Domain:** CGP Grey 48–49, KINKISHARYO 50-51, Le Petit Journal 56–57, Siemens Pressebild/permission Siemens Germany/Chrisitian Kuhna 42–43; **Shutterstock:** 4 PM production 30-31, Tim Arbaev 41b, ArtWell 34, Mawardi Bahar 58–59, biletskihevgeniy.com 11, Brandon Blinkenberg 41t, Anucha Chechung 26–27, cyo bo 52–53, D-Visions 1, 24–25, dragon fang 45, Frame Stock Footage 22–23, JirikV 28, Matej Kastelic 54-55, Gonul Kokal 20, Peter Kotoff 18-19, Anton Kozolovsky 29, TK Kurikawa 38–39, Rawpixel.com 12-13, Richard Semik 49, Dmytor Shestakov 14–15, ShutterStockStudio 6–7.

STEM

Key Concepts in STEM: Physics describes the key theories in the study of physics and their practical applications. This book, *Magnetism*, describes the properties of magnets and their uses. Diagrams and informative photographs, detailed features on related aspects of the topics covered and the main scientists involved in the advancement of physics, and definitions of key "Science Words" all enhance the coverage. "Try This" sidebars outline experiments that learners can follow as a first step to practical investigations.

Publisher's note to educators and parents: Our editors have carefully reviewed the web sites included at the end of this book to ensure that they are suitable for students. Many web sites change frequently, however, and we cannot guarantee that a site's future contents will continue to meet our high standards of quality and educational value. Be advised that students should be closely supervised whenever they access the Internet.

Contents

MAGNETIC MATERIALS

A magnetic rock called lodestone has a power that seemed magical to early people. It can point to the north and guide a traveler even when it is dark, or in bad weather. Early explorers used this "magic" on their sea voyages.

Lodestone, or magnetite, was first recognized by Chinese scholars as early as 2700 BCE. It contains iron and attracts other metals. Pieces of lodestone will either attract or repel each other, depending on what they are being used for. Lodestone demonstrates another amazing property when it is set up so that it can move freely. It will turn until it is pointing roughly north–south. This magnetic "compass" may have been used by navigators as early as the 3rd century CE in China. Certainly it was in widespread use by European navigators in the 12th century.

A magnetic needle enabled this portable sundial to be aligned north–south to take a correct reading. The instrument doubled as a magnetic compass.

MAGNETIC DOMAINS

Tiny areas in a piece of iron, called domains, are like individual magnets. Applying a magnetic field lines up these magnets. The metal is said to be saturated when all the domains have their magnetism lined up.

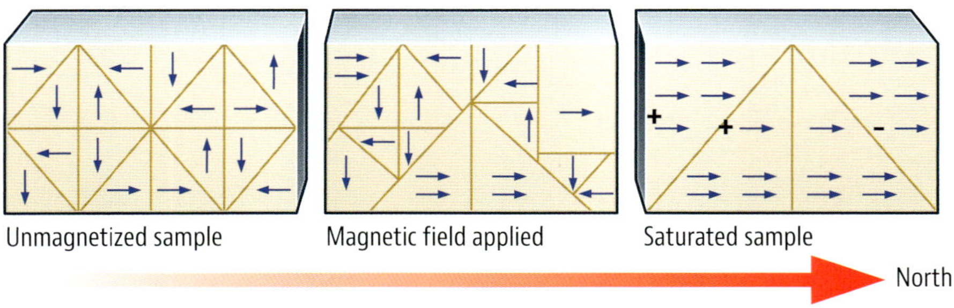

Making Compasses

Supplies of lodestone were limited. But it was possible to use a piece of lodestone to make a large number of magnetic needles for use in compasses. By stroking an iron needle with the lodestone many times, always in the same direction, the iron became magnetized.

In the 20th century, this effect was explained by the French physicist Pierre-Ernest Weiss (1865–1940). Individual atoms of iron and other elements present in a piece of metal are like miniature magnets. The metal is divided into regions called domains in which all the atoms line up. The magnetism of the atoms adds together, making each domain into a tiny magnet. But the atoms in different domains point in different directions, so that overall the piece of metal does not behave like a magnet. Repeated stroking with a piece of lodestone causes the atoms to turn.

Eventually all the atoms point in the same direction, and their magnetism adds up to make the whole piece of metal a strong magnet. When all the atoms are lined up and the magnetism cannot become any greater, the piece of metal is said to be saturated.

SCIENCE WORDS

Atom: The smallest part of a chemical element that can exist on its own. It has a central nucleus, surrounded by electrons.
Domain: A small region in a magnetic material in which the magnetic fields of individual atoms all point in the same direction, making the domain into a single small magnet. In the unmagnetized material, fields of different domains point in different directions, canceling one another out.
Lodestone: A naturally magnetic iron ore, formerly used to make magnetic compasses. It is one of only two minerals that is naturally magnetic; the other, pyrrhotite, has only weak magnetism.

TRY THIS

How Much Pull?

Magnets vary in strength, usually depending on how large they are. One way to measure the strength of a magnet is to see how much it will lift up.

What to do

Using strong masking tape, tape a bar magnet securely to the edge of a table, with one end sticking out over the edge. Add paper clips to that end, building up a chain of clips hanging down. What is the maximum number of clips the magnet will keep up? Repeat the experiment, this time using a horseshoe magnet. How many clips will this magnet hold up?

The first paper clip is held by magnetism. Paper clips are made from steel, a magnetic material. In contact with the magnet, a paper clip also becomes a magnet. The new paper-clip magnet holds up the second clip. That becomes a magnet, and so on. Generally, a horseshoe magnet is stronger than a bar magnet of the same length.

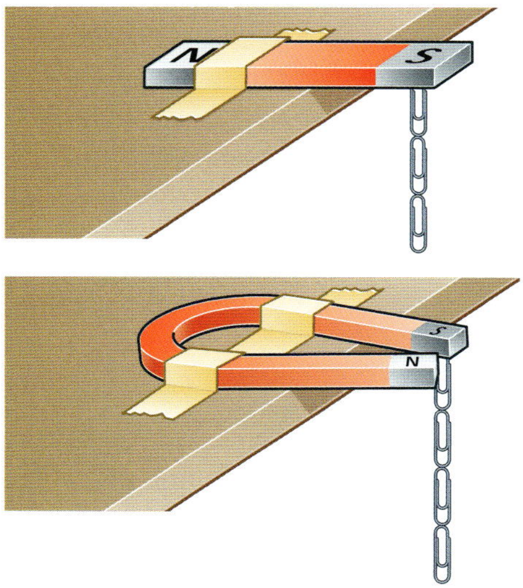

Test the strength of a magnet by seeing how many paper clips it will hold up.

MAGNETIC POLES AND FIELDS

A magnet has a "sphere of influence" called a magnetic field that surrounds it and extends into nearby space. How another magnet is affected by the first magnet depends both on this magnetic field and on the second magnet's own properties.

A piece of lodestone or other magnet will attract small metal objects to itself. The way they arrange themselves around the magnet reveals important facts about magnetism. Iron filings—tiny metal grains produced when a block of iron is filed or machined—show this especially clearly. They cluster around the two ends of a bar magnet, but not toward its middle. If the magnet is in the shape of a horseshoe, the filings cluster at the ends of the two "legs," but not around its middle.

ONE OR TWO POLES

These fields are produced by one or two magnetic poles. Such poles always occur in linked pairs, but if they are widely separated they can be regarded as single poles.

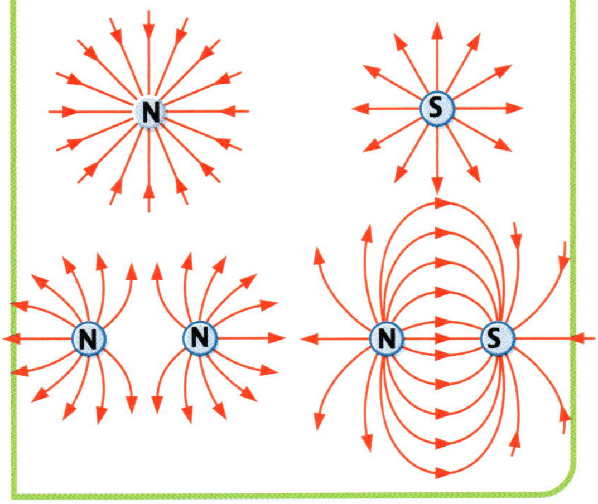

Every magnet has these two "attractive ends," called its poles. One of the poles of a magnet is attracted to the north and the other to the south. This is the reason why a compass needle swings until it is pointing north–south. The north-seeking pole is called the magnet's north, or N, pole; the south-seeking pole is the south, or S, pole.

Attraction of Opposites

When the two north poles of a magnet are brought close to each other, they can be felt to push away or repel each other. The closer together the two north poles are, the harder is the push. The same happens if the two south poles are brought close to each other. But when a north pole is brought close to a south pole, a force of pull or attraction can be felt. Again, the force is stronger the closer the poles are to each other.

Iron filings scattered around the poles of a bar magnet reveal the magnetic field around them. Each filing moves in the direction of the field at its location, and together they trace a dramatic picture of the whole magnetic field.

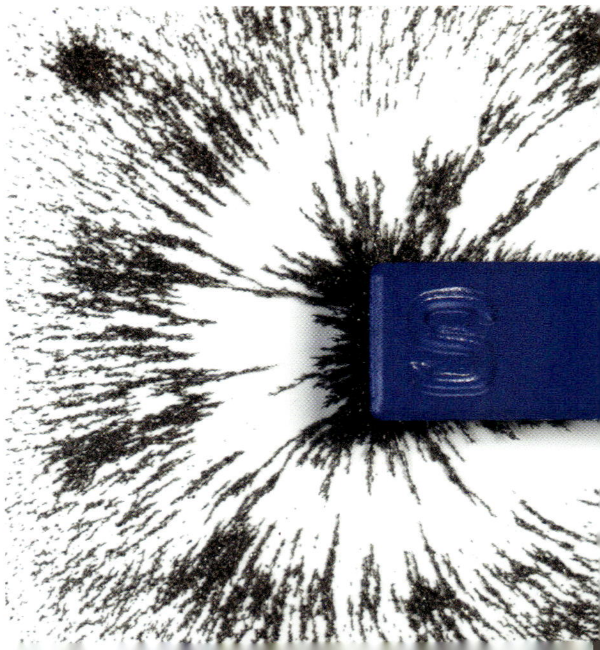

PLOTTING THE FIELD

Trails of dots marking successive positions of the head and tail of the compass needle trace out magnetic field lines.

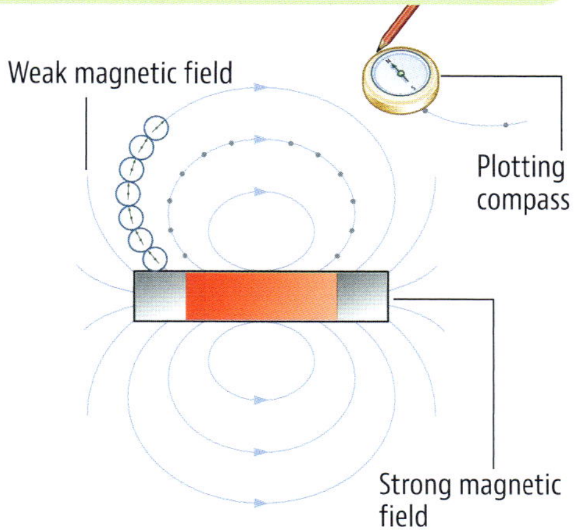

The field can be "mapped" using a small compass. If it is brought close to a magnet, the needle of the compass points toward or away from the nearest pole of the magnet. If the nearest pole is a south pole, the north-seeking end of the needle points toward it.

FORCES BETWEEN MAGNETS

Two bar magnets suspended on strings will push each other apart or pull each other together depending on whether like poles (two N or two S) or unlike poles (N and S) are closest to each other.

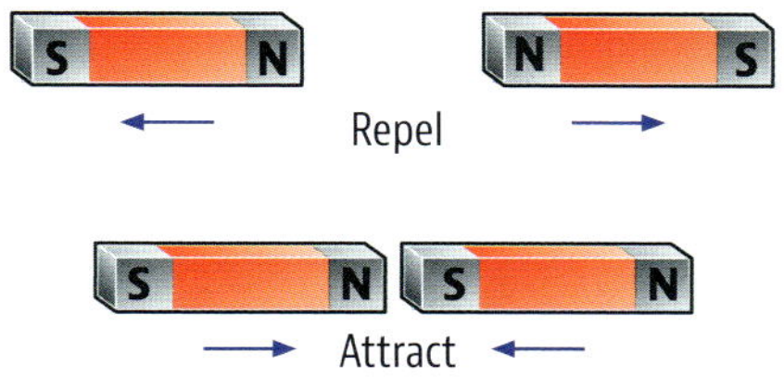

If the compass is moved toward the other end of the magnet, the effect of the north pole gradually increases, and the needle turns until its south-seeking end points toward the north pole.

To map the field, put the magnet on a piece of paper and place the compass anywhere on the paper. Mark the two ends of the needle with pencil dots. Then move the compass until the tail of the needle is just over the dot that marks the previous position of the head of the needle. Mark the new position of the head of the needle, and again move the compass until the tail of the needle is at that position. Continue repeating this process. The result is a chain of dots that can be joined up smoothly to make a curved line. At every point along this line the direction of the line is the direction of the compass at that point, which is the direction of the force that would be experienced by any magnetic object at that point. The curved line is called a line of force or a field line. The region in which magnetic forces are experienced is the magnetic field. Field lines indicate the north–south direction.

TRY THIS

Field of Force

The region around a magnet in which it attracts metal objects is called its magnetic field. The field is invisible, yet it can be shown. This project shows you how to reveal the existence of a magnetic field and discover its shape.

What to do

Put some iron filings in a cup and sprinkle them thinly onto a piece of writing paper. Carefully slide the paper into a clear plastic pocket and tape the open edge of the pocket closed (this keeps the filings from spilling and possibly sticking to the magnet). Put the magnet on a table and place the plastic pocket on top of it. The filings will arrange themselves in a pattern that corresponds to the magnetic field of the magnet. Repeat the process with another shape of magnet if you have one. You can even use the same method to find the shape of the magnetic field between two magnets.

The iron filings line themselves up along the lines of force that make up the magnetic field. The field is strongest near the poles (ends) of the magnet, which is why the filings concentrate at these points. If you use two magnets, you will see how their magnetic fields combine.

The iron filings inside the plastic pocket will reveal the magnetic field.

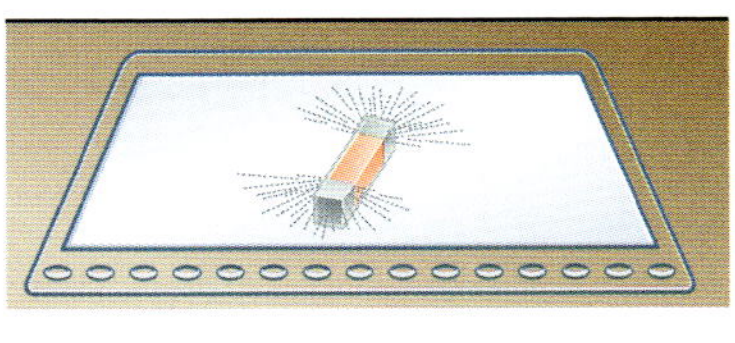

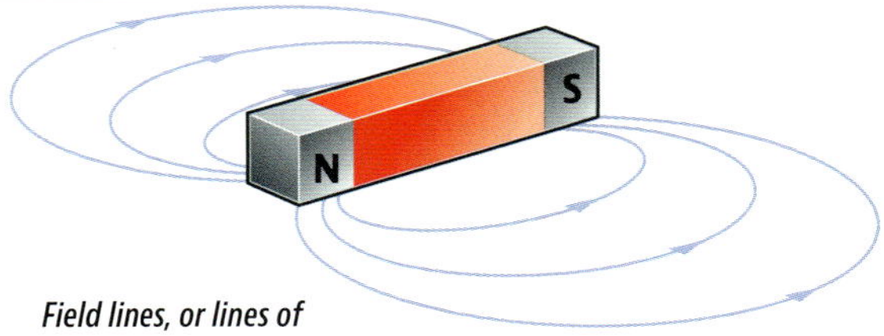

Field lines, or lines of force, are imaginary lines showing the force that the magnet would exert on an imaginary north magnetic pole. At each point the direction of the field line shows the direction of the force.

Picturing the Field

The magnetic field can be made visible in a striking way with the aid of iron filings. If a sheet of paper is placed over a magnet and iron filings are shaken over the paper, many cluster directly over the poles. Some lie scattered over the rest of the area around the magnet and arrange themselves in curved lines. The lines run from one pole to the other. They are formed as each individual grain lines up in a certain direction. Each grain acts like a tiny magnetic compass and points in a definite direction. Together they give a vivid picture of the form of the field, and even of its strength. The lines crowd together near poles where the field is strongest.

The direction of the field at a point is defined as the direction of the force that would be experienced by the north pole of another magnet at that point. According to this definition, the field lines run outside the magnet from its north pole toward its south pole. Entering the magnet near the south pole, they return toward the north pole, each forming a closed loop. Field lines do not meet or cross one another.

Poles Apart

Every magnet has two opposite poles: no isolated pole is known to exist. (Some physicists continue to search for subatomic particles that are single magnetic poles, but many doubt that such "monopoles" will ever be discovered.) This makes working out the laws of the magnetic field difficult. It is easier to see what is happening with a very long bar magnet, with the poles far apart. The magnetic effects of each pole can be studied while the influence of the other one is ignored.

The strength of the magnetic field produced by a single pole can be measured at each point in space by measuring the force on another magnetic pole placed there. It is found that the strength of the field falls to one-fourth when the distance is doubled, to one-ninth when it is tripled, to one-sixteenth when it is increased fourfold, and so on. That is, the field strength is inversely proportional to the square of the distance. Many other physical quantities also fall off in strength proportional to the square of the distance, including the strength of gravity and of electrical fields, and the brightness of light.

The strength of the magnetic field at a given point depends not only on how far from the magnet that point is, but also on the strength of the magnet. A particular piece of iron can be more or less strongly magnetized. (The word "magnetic" describes a material that can be made into a magnet. The word "magnetized" refers to a magnetic material that actually is a magnet.) Magnetic materials include iron, cobalt, and nickel, neodymium alloys, and steel. Scientists have also created nonmetallic materials called ferrites that can be made into powerful magnets.

SCIENCE WORDS

Bar magnet: A permanent magnet that is bar-shaped, with a magnetic pole at each end.
Field: The pattern of magnetic, electric, gravitational, or other influence around an object. See also Line of force.
Line of force: An imaginary line where the direction at any point in a field shows the direction of the field at that point.
Pole: One of the two points of a magnet where the strength of the field is greatest. Lines of force diverge (radiate out) from one pole and converge (meet at a point) on the other.

A FORCE BETWEEN POLES

Two north poles repel each other, as do two south poles. A north and a south pole attract each other. The force between two magnetic poles is proportional to M_1 and M_2, the strengths of the magnets, and inversely proportional to d^2, the square of the distance between them.

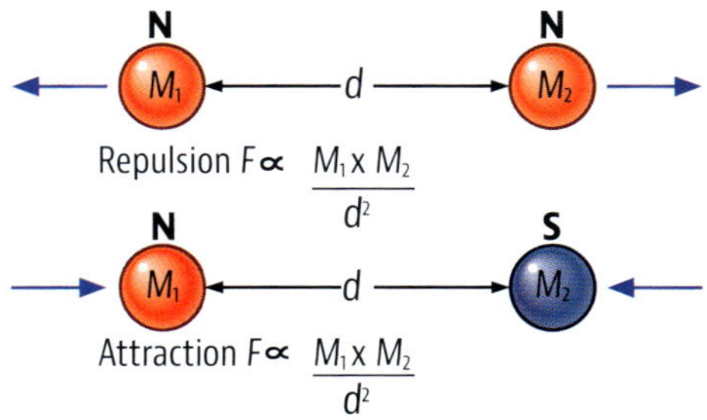

FIELDS BETWEEN MAGNETS

When two bar magnets are brought near to each other, the shapes of the field lines between them alter depending on which poles are involved.

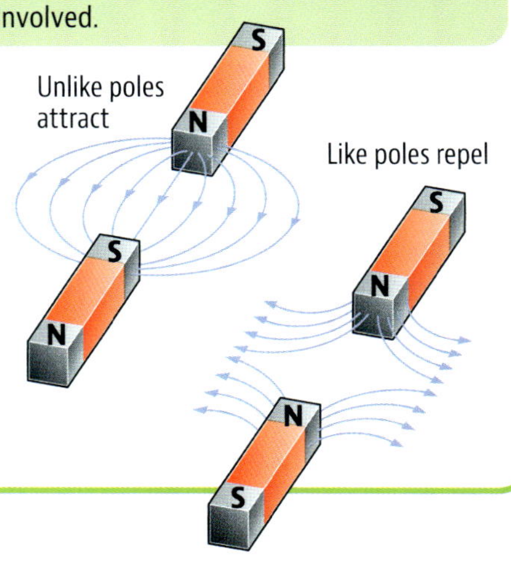

EARTH'S MAGNETISM

Earth is a giant magnet. The turning of Earth's core generates the planet's magnetic field, which not only influences compasses, but also reaches far into space. It changes slowly with time, and the record of these changes is captured in the rocks, revealing the planet's geological history billions of years back into the past.

Earth's magnetic field is like a huge bar magnet deep within the planet quite close to Earth's axis of rotation. The poles of this imaginary magnet lie beneath two opposite points on Earth's surface, close to the geographical poles, where the field lines are vertical. The "magnet" inside Earth has its south pole near Earth's north pole, and the north pole near Earth's south pole.

SCIENCE WORDS

Magnetic compass: A device in which a freely rotating magnetized needle indicates the direction of magnetic north.
Magnetic equator: The imaginary circle around Earth, approximately halfway between the planet's magnetic poles, where Earth's magnetic field is horizontal.
Magnetic north: The direction in which a magnetic compass points (in the northern hemisphere). Except in the polar regions, this direction is close enough to the true north–south line to be valuable to navigators. Near the poles, the best way of determining position accurately is celestial navigation.

Shifting, multicolored auroras are often seen in the sky in polar regions. They are caused by electrically charged particles, mostly from the Sun, being funneled by Earth's magnetic field down from space toward the magnetic poles.

EARTH AS A MAGNET

The lines of force of Earth's magnetic field can be traced just as they can for a small magnet. The field has the same shape as the magnetic field of a bar magnet. The north magnetic pole lies about 800 miles (1,300 km) away from the geographical north pole. The south magnetic pole is about twice as far as this from the Earth's geographical south pole.

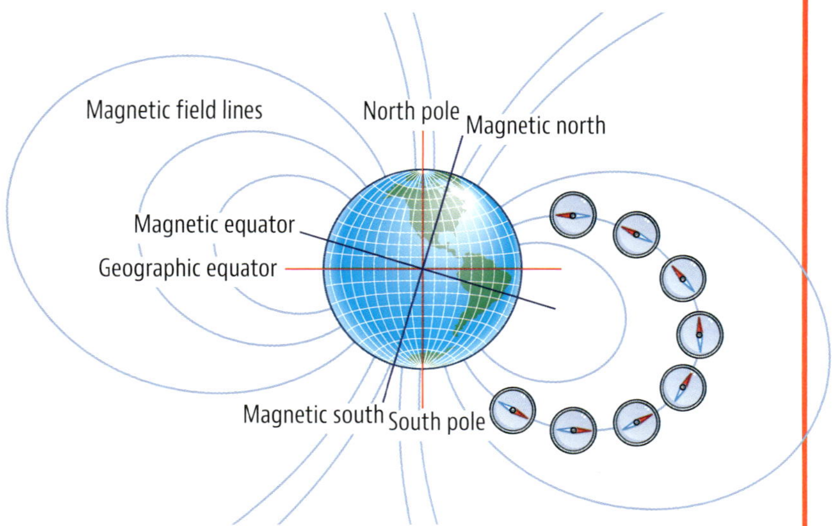

ANGLE OF DIP

A magnetic needle dips because it tends to point toward the nearest pole of the imaginary magnet deep within Earth.

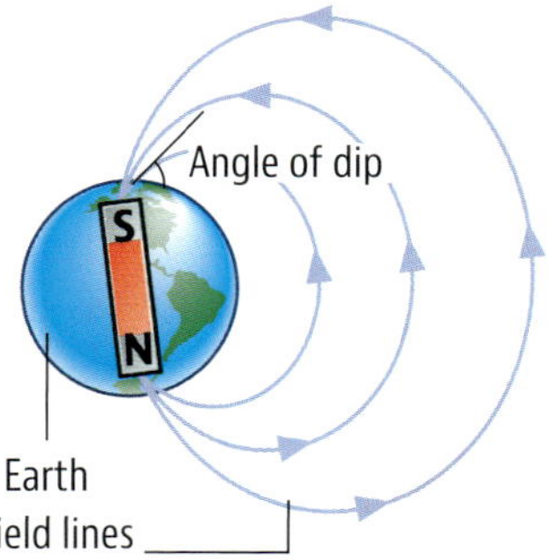

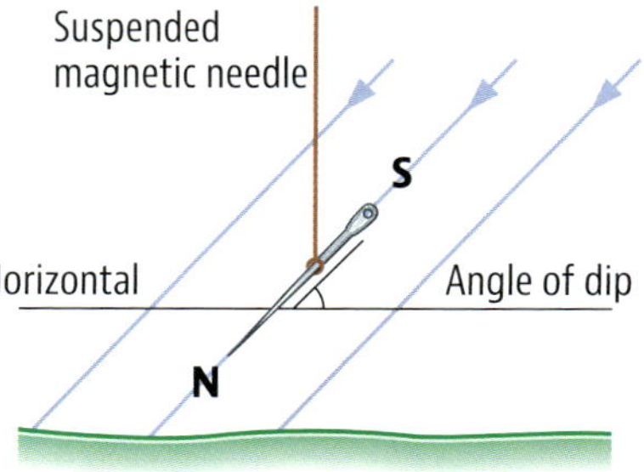

The points on the surface of Earth are called Earth's north and south magnetic poles. The magnetic equator is the imaginary line around Earth, halfway between the magnetic poles, where the field lines are horizontal.

Because a magnetic compass points to the magnetic poles, it does not in general point to true north or true south. The angle between true north and "magnetic north" is called variation. The direction of magnetic north is shown on some maps. Variation changes with time because the positions of the poles change: they wander by about 12 miles (20 km) per year.

Animal Navigators

Many types of animal rely, at least partly, on Earth's magnetism to guide them. Some birds migrate from one continent to another at the beginning of summer and winter. They get their sense of direction largely from the position of the Sun or stars; but when the skies are cloudy and they cannot do that, their sense of Earth's magnetism guides them. Water-dwelling organisms that are sensitive to Earth's magnetic field include some bacteria, whales, dolphins, sharks, and sea turtles.

The Magnetosphere

Earth's magnetic field extends into space. Electrically charged particles that constantly flow outward from the Sun are called the solar wind. They are trapped by Earth's magnetic field and form two thick belts of charged particles, known as the Van Allen belts. Earth's magnetosphere is the volume

SCIENCE WORDS

Magnetosphere: The region around Earth or another celestial body, such as Jupiter, in which its magnetic field is stronger than the field in surrounding space.

Solar wind: The stream of charged particles that are ejected from the Sun's upper atmosphere.

Van Allen belts: Another name for Earth's radiation belts.

of space in which its magnetic field is stronger than that of the Sun. The solar wind squeezes the magnetosphere, so that it is 40,000 miles (64,000 km) above the surface on the sunward side of Earth, but stretches into a tail, over 600,000 miles (1 million km) long, on the side away from the Sun. Some other planets, in particular Jupiter, have magnetic fields and magnetospheres. The Sun's magnetic field is powerful and dominates an area of space, called the heliosphere, which extends beyond the distant dwarf planet Pluto.

Restless Planet

Although lodestone is the only strongly magnetized rock, many other types of rock were weakly magnetized when they were first formed. The direction of their magnetic field was the same as that of Earth at that time, but since then the rocks have moved.

By patient detective work it is possible for scientists to reconstruct how the rocks have moved since their formation. This work shows that the continents have drifted over the surface of Earth ever since a single supercontinent called Pangaea broke up about 180 million years ago. Electric currents in Earth's molten outer core generate its magnetic field.

In addition to the slow changes in direction of the field that cause "polar wandering," every few hundred thousand years the field dies down to zero and then increases in strength again, but pointing in the opposite direction. The record of these field reversals is recorded in the magnetism of the rocks.

Like the Aurora borealis (North) and Aurora australis (South) on Earth, Jupiter's poles attract charged particles from the Sun. This is Jupiter's aurora around the planet's north pole, photographed by NASA's Hubble Space Telescope.

TRY THIS

Make a Compass

A compass consists of a magnetized needle that is free to turn and point in any direction. But in Earth's magnetic field one end of the needle always points toward north. In this project, make a simple floating compass.

What to do

Fill a shallow bowl three-fourths full of water. Tape a sewing needle across a slice of cork or Styrofoam. Using one end of a magnet, stroke the needle 20 or 30 times along its length, always in the same direction. Float the cork or foam with its needle in the center of the water. After a while the needle will settle down and point north–south.

The magnetized needle lines up with the direction of the Earth's magnetic field. Because this direction is north–south, the needle also points north–south. You can change the needle from its north–south alignment by bringing a magnet near it. The needle lines up with the strong field of the magnet rather than the weak field of Earth.

The needle on the compass will settle pointing north–south.

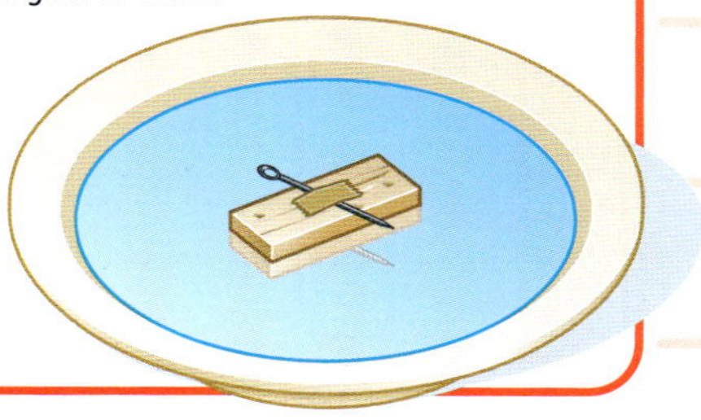

MEASURING AND DETECTING MAGNETIC FIELDS

Compasses are the oldest science tools that make use of Earth's magnetic field. Earth behaves as if it has a giant bar magnet along its axis between the north and south poles.

In a magnetic compass, the pointer or needle is itself a small magnet, pivoted at the center so it can swing from side to side. When it points to the north, it indicates the direction of Earth's magnetic field.

A compass is not the only scientific instrument that uses magnets. A device called a tangent galvanometer detects electric currents by using a small pivoted magnet. Fairly similar in this respect is a magnetometer, which, as its name suggests, is an instrument for measuring the strength of a magnetic field. The magnetometer consists of a small pivoted magnet attached at right angles to a long, nonmagnetic pointer.

MAGNETOMETER

A magnetometer consists of a small pivoted magnet that is attached at right angles to a long pointer that indicates the direction of the magnetic field **(a)** of Earth and **(b)** of a nearby magnet.

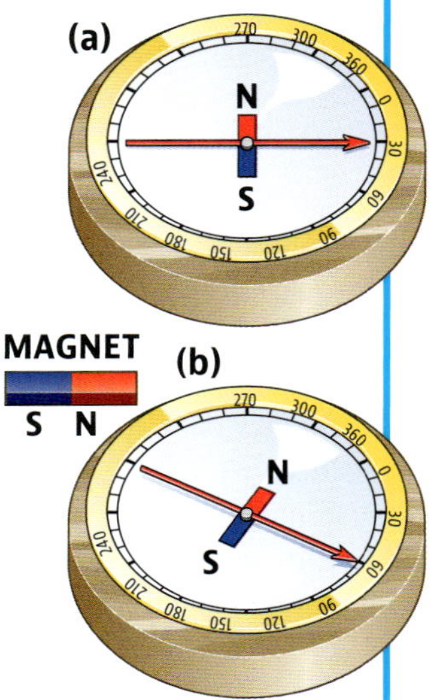

The device is turned until the magnet lines up with Earth's magnetic field, and the pointer indicates zero (illustration **a**, left). When another magnetic field, such as that of a bar magnet, is nearby, one end of the magnet is attracted to it because opposite magnetic poles, a north and a south, attract each other. This makes the pointer move around the scale (illustration **b**, left). The angle indicated by the pointer is a measure of the strength of the external field. The actual field strength is proportional to the tangent of the indicated angle.

Just as opposite magnetic poles attract each other, similar ones—two north poles or two south poles—repel each other. This effect is put to good use in a maglev train. Maglev is short for "magnetic levitation" (lifting); electromagnets with one polarity inside the train lift the vehicle clear of rails that have the opposite polarity. Electromagnetic effects like this are used in a wide range of devices,

A metal detector works by creating a magnetic field and responding to changes in the field caused by the presence of metallic objects. Using such devices, people can unearth metal objects buried below the ground.

including electric bells, relays, loudspeakers, generators, and electric motors.

Magnetic Detectors

The presence of nearby metal objects can alter a magnetic field. That is what makes metal detectors work—you may have seen them in airports or stores. The apparatus consists of a tall archway the size of a door. Large coils of wire in the sides of the arch carry an electric current, which creates a magnetic field. Anyone who walks through the arch carrying a metal object, such as a handgun, will disturb the magnetic field. The change in the field is detected and sets off a warning alarm. A bunch of keys or a metal watchband will also activate the alarm, which is why people are asked to remove these objects before they walk through the detector at airports.

SCIENCE WORDS

Maglev: Abbreviation for "magnetic levitation," the technology for keeping vehicles such as trains seemingly floating above the track on which they run by means of mutually repelling magnetic fields. This reduces friction.

Treasure Hunters

People searching for buried objects use a different type of metal detector. It consists of a coil of wire carrying an electric current mounted inside a disk at the end of a long handle. Batteries in the handle supply the electricity. The coil produces a magnetic field that is distorted by any metal objects in the ground. A microchip in the handle detects the tiny changes in current produced by the field distortion and causes the detector to make a buzzing or bleeping sound. Scientific versions of these instruments give an audible warning through headphones.

Magnetometer and Minerals

Highly sensitive magnetometers are used in prospecting for underground minerals. To cover a large area quickly, the instrument is mounted in an aerodynamic "torpedo" and towed behind a light airplane or helicopter. The device works by detecting distortions in Earth's magnetic field that have been caused by mineral deposits. Minerals that are nearer the surface have a greater effect on the magnetometer than those minerals buried deep underground. The data from the magnetometer passes along a cable to an onboard computer in the towing aircraft.

MAKING MAGNETS

Magnets were precious objects in centuries past when explorers relied on them for navigation. Later, scientists learned how to make magnets from iron and steel. Today, there are effective ways of making strong magnets for the devices we all rely on.

It was probably in ancient times when people first discovered that hammering an iron bar while it was aligned north–south magnetized it. If the iron is made red-hot and then allowed to cool while being hammered, it becomes even more strongly magnetized. Stroking a piece of iron with a magnet also magnetizes it. Sailors used to take a piece of lodestone on voyages so that they could "refresh" their compass needles or magnetize new ones whenever necessary. Hammering an iron bar is a good substitute for the lodestone (see below).

These methods of magnetizing the metal depend on the fact that a piece of unmagnetized iron or steel is made up of

HOW TO MAKE A MAGNET

1 Using another magnet
A magnet can be made from a steel bar by stroking it repeatedly with a bar magnet. In unmagnetized steel, the magnetic domains point in all directions. Their north and south poles cancel each other out. Stroking the steel with a bar magnet pulls the domains around to point in the same direction, magnetizing the steel.

2 Hammering in a magnetic field
Hammering an iron bar gradually magnetizes it. The domains in the iron bar are jumbled to begin with. Each hammer blow gradually turns them, pulled by the Earth's field (which is nearly vertical in polar regions). Finally, the domains line up parallel to the Earth's field, forming a magnet.

3 Using a coil
Applying a magnetic field generated by an electric current can also magnetize an iron bar. The domains remain lined up, although possibly only weakly, after the current is turned off.

1

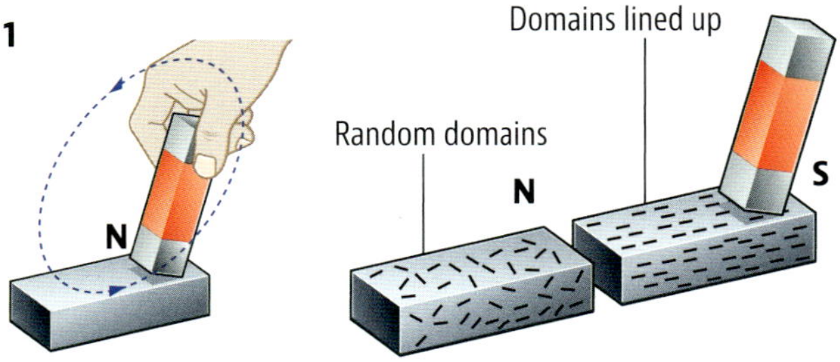

2

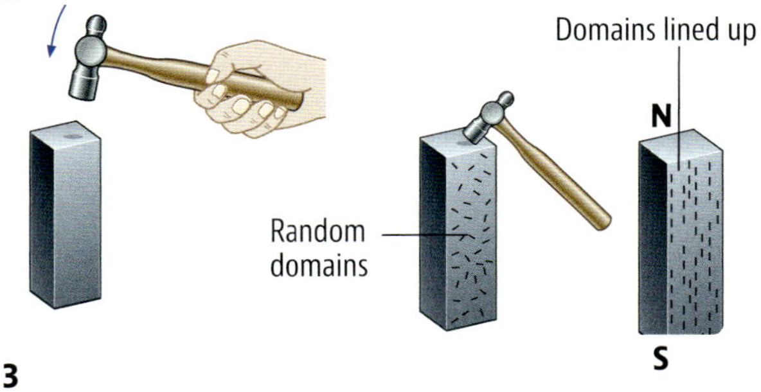

3

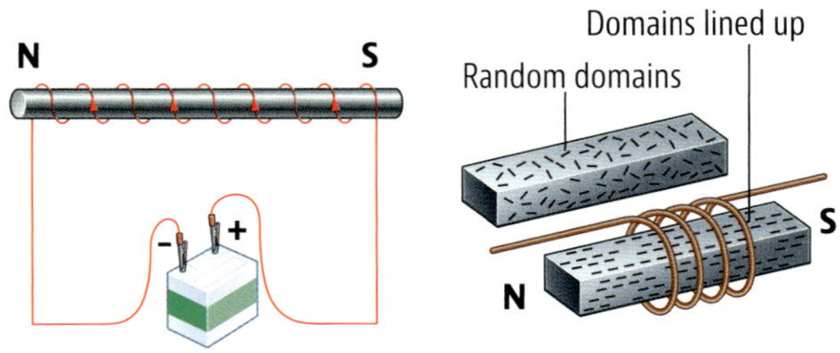

countless tiny magnets—the domains or regions where atoms are all aligned (see pages 4 and 5). Because they are pointing in all directions, these domains cancel one another out, and the metal has no overall magnetization. To turn the metal into a magnet, the atoms in all the domains have to be rotated so that they all point in the same way. An extremely strong magnetic field can do this. A coil of wire carrying an electric current will provide such a field. The field of another magnet that is made to stroke the piece of metal repeatedly will also magnetize the metal. If the north pole of the magnet makes contact with the metal being magnetized, it will pull the south poles of the atomic magnets toward itself. A south pole will develop at the end of the metal where the magnet breaks contact.

Shaking up the Atoms

Earth's magnetic field is strong enough to align the atoms and magnetize the metal if the atoms are constantly disturbed by hammer blows or by being heated. Heat makes a metal's atoms vibrate more vigorously. However, the opposite is also true. Heating a magnet or repeatedly striking it with a hammer "shakes up" the atomic magnets. They take on a random arrangement again, and the magnet will lose its magnetism.

Objects that are left for a long period of time near a magnet will become magnetized. The loudspeakers of audio equipment contain strong permanent magnets. Small metal objects such as screws can become magnetized if they are left for any length of time near speakers. From the early 1970s through to the 1990s, sound was recorded as a pattern in the metal coating of a plastic audiotape. If these audiotapes were left next to speakers they were spoiled because the magnetization was altered. Stray magnetic fields can affect other types of magnetic storage, but USB drives and memory sticks are not made from such magnetic materials so cannot be damaged by magnets. This type of storage media is called solid-state memory.

TRY THIS

Captive Filings

Once a piece of iron or steel is magnetized, it usually stays magnetized. It becomes a permanent magnet, and its magnetic field cannot be turned on and off. However, an electromagnet can be turned on and off. In this project, you will make some iron filings jump using an electromagnet.

What to do

Use the point of a nail to punch a hole through the center of a piece of 6 in (15 cm) square thick card. Wind a length of wire around the nail, leaving about 1 ft (30 cm) of wire free at each end. Thread one end of the wire through the hole in the card, push the nail through and stand it on a reel of tape so that the card is flat and the nail is upright (see below). Sprinkle iron filings thinly on the card around the nail. Connect the ends of the wire to the terminals of a battery, and watch what happens to the iron filings. They will form the pattern of a star around the nail, which is acting as an electromagnet. Do not leave the battery connected for long as the wires and the nail will get hot. When you disconnect one wire, the filings will no longer be attracted by the electromagnet and can be shuffled about. As soon as the wire is reconnected to the battery, the filings will jump back into their starlike pattern.

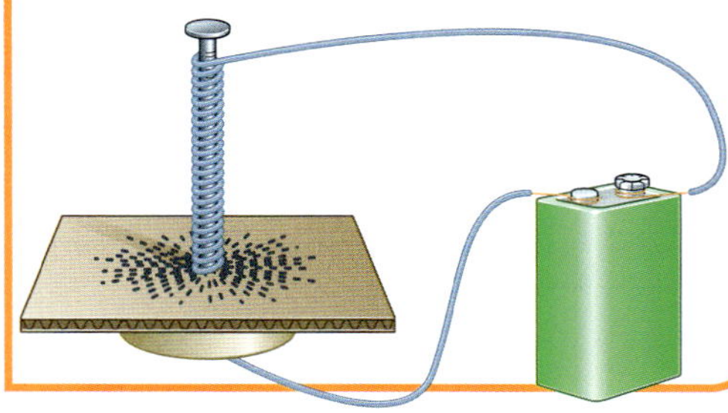

USES OF MAGNETS

Magnets, often working hand-in-hand with electricity, are used in radios, loudspeakers, microwaves, disk drives, magnetic tape, and permanent magnet motors. They are widely used in industry, while in modern medicine they are an essential tool for revealing the workings of the living human body.

Magnets are indispensable to modern life. One important use is in computers. Most personal computers still have hard drives, although increasingly they are switching to solid-state drives (SSDs). Devices such as smartphones, tablets, and some laptops now only use SSDs. Computer hard disk drives are coated with magnetic material. Data is stored in this coating by a moving arm called a read–write head, which applies a magnetic field to

A computer's hard disk drive contains two or more rigid metal platters coated with a magnetic material. Read–Write heads swing back and forth across each platter as it spins rapidly.

MAGNETIC STORAGE

A kind of magnetic storage that is built into most personal computers (PCs) is the hard drive. This uses one or more hard disks. Until the late 1990s, most PCs also incorporated a "floppy" disk drive, using a single flexible magnetic disk. These have now been replaced by USB flash drives, CD-ROM, and DVD drives, which are nonmagnetic.

Old-fashioned computer systems used huge banks of magnetic tapes like these.

the disk to make a tiny magnetized area, corresponding to a bit. A bit is a 0 or a 1, which are the basic units of information used in connection with computers and digital communications. The read–write head can also read any information that is already on the disk by detecting the magnetization of the surface directly beneath the head.

In the past, some large mainframe computers, which were needed to store the vast quantities of information required by major corporations, universities, or the military, used magnetic tapes. These consisted of plastic film with a coating of a magnetic material. The same principle was also used for the audiotapes and videotapes that were used for storing information that represented sounds and pictures. A magnetic signal was recorded onto a magnetic disk or tape by applying a fluctuating magnetic field. A varying electric current flowing through an electromagnet in the recording head created this field. When the magnetic signal was read, the fluctuations in the magnetism of the

TRY THIS

How Attractive?

A magnet attracts some materials, but not others. In this project, various materials are tested to see which are best attracted to the magnet.

What to do

Collect these different objects to test: a pin, coins, an eraser, a piece of aluminum foil, a wax crayon, and a thumbtack. Touch each object in turn with the magnet. Note down which of the objects are attracted to the magnet. You could put those that are attracted in one pile and those that are not in another pile. Do the attracted objects have anything in common?

If you look at the attracted objects, you will see that they are all made from steel. Objects made from other metals, such as copper coins, nickels, and aluminum foil, were not attracted. So it is only steel objects that are attracted to the magnet. These objects themselves become magnetized. Notice that none of the nonmetallic objects are attracted by the magnet.

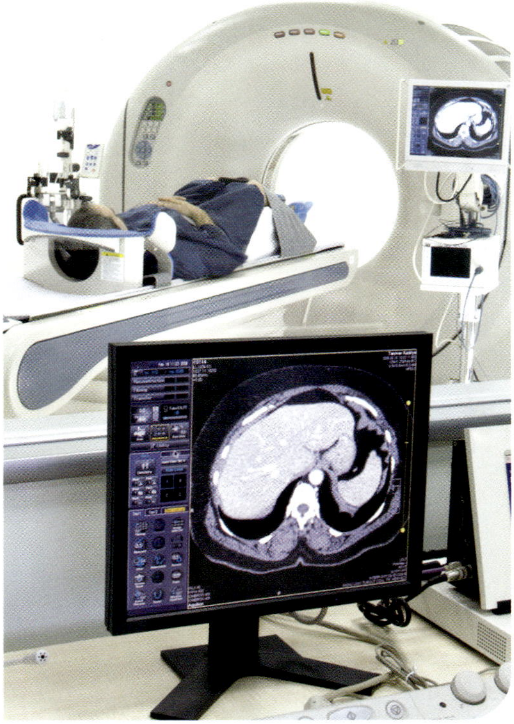

An MRI scanner uses a powerful magnetic field and radio waves to make the hydrogen atoms in the patient's body "broadcast" their own radio signals, from which a computer can build up an image of the patient's internal organs.

recording surface as it sped past the detector generated a varying electric current in the detector. This current was then amplified to give the output. Although this type of magnetic tape is still used, it is rarely used outside of professional sound or video recording production studios.

Magnetic Pictures

Modern medicine relies heavily on the ability to create images of organs inside the living human body. There are several techniques for doing this, and one of the most important is called magnetic resonance imaging (MRI). When a patient has an MRI scan, they lie down and are positioned with the relevant part of the body, such as the head or chest, in the scanner. A powerful magnetic field, tens of thousands of times as strong as Earth's field, is turned on. The patient's body is then bathed in radio waves. The magnetic field puts the hydrogen atoms in the patient's tissues into a state in which they can absorb energy from the radio waves. When the signals are turned off, these hydrogen atoms reradiate the energy they have absorbed. Detectors pick up these "broadcasts," and a powerful computer assembles the signals into a colored image of the interior of the body. The brightness and color of each part of the image represent the amount and type of tissue present at the corresponding point in the body. A technique called functional MRI (fMRI) can show the activity going on in a person's internal organs —for example, the activity of the parts of the brain involved in emotional reactions or in tasks such as mental arithmetic.

Heavy Lifting

Electromagnets are important in industry. They are magnets in which the magnetism

SCIENCE WORDS

Electromagnet: A device that develops a magnetic field when an electric current is passed through it. It consists of a coil with a core.

Magnetic resonance imaging (MRI): The technique of making pictures of the inside of the living human body. A person is placed in a strong magnetic field and bathed in radio waves. Hydrogen atoms in the body "rebroadcast" the waves.

Superconductivity: The property of conducting electricity with no resistance at all. Some metals do this when cooled to a temperature close to absolute zero (–459.67°F/–273.15°C).

is generated by electric current. The magnetic field can be much stronger than is possible in solid permanent magnets, and it can be switched on and off simply by switching the current on and off. Strong electromagnets are used in cranes for lifting iron and steel objects, such as junked vehicles at wrecking yards. They can also be used for separating metal ores and scrap metal. This is an essential preliminary to processing the material to get a high level of purity in the metal that is refined from the ore or recovered from the scrap. The materials to be separated are dropped onto a drum rotating around a semicircular electromagnet. Any nonmagnetic materials drop off into a bin. The drum holds magnetic materials as it rotates until they pass the magnet, when they fall into a separate container.

MAGNETIC SEPARATOR

Magnetic and nonmagnetic scrap metals can be separated from each other by means of a rotating drum and an electromagnet. Copper and brass are among the metals that simply slide off the drum. Iron, steel, and other magnetic metals stick to its surface until they pass the magnet and fall off.

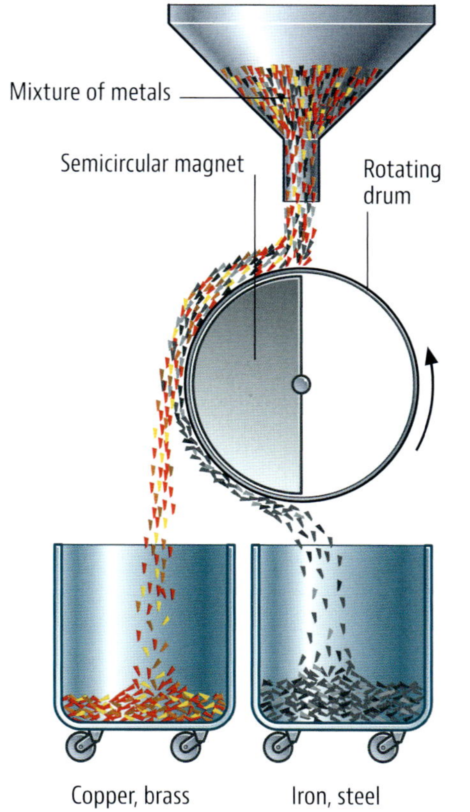

Guiding Particles

Scientists learn about the structure of matter by accelerating subatomic particles in giant machines called colliders and then smashing the particles into one another. The Large Hadron Collider (LHC) is the world's largest and most powerful particle accelerator. It is housed in an enormous circular underground tunnel 17 miles (27 km) in circumference at CERN, the European Laboratory for Particle Physics, near Geneva in Switzerland.

The particles, which are traveling at almost the speed of light, are held on course on the curved path by electromagnets that encircle the path of the particle beams. The electromagnets have to be very powerful, so superconductors are used. These are metals kept at low temperatures so that they become superconducting—meaning they can pass electric currents without any electrical resistance and therefore lose almost no energy as wasted heat. Magnetic fields are also important in the detectors that record the paths of the particles produced in the collisions. A magnetic field is generated in a particle detector to bend the paths of the particles produced. The amount of curvature depends on the speed of the particle and on its mass. Measuring the curvatures of the paths that appear in the pictures provides important data on the particles.

ELECTRICITY AND MAGNETISM

For nearly 200 years, scientists have known that electricity and magnetism are closely interlinked. Electric currents generate magnetic fields; and magnetic fields, when they alter, can drive electric currents. Scientists now regard electricity and magnetism as two aspects of the same thing—electromagnetism.

The link between electricity and magnetism was first discovered quite early in the 19th century. The Danish physicist Hans Christian Ørsted (1777–1851) found that a wire carrying an electric current will affect a magnetic needle. The magnetic field lines around a wire are circles. If you look along a wire, and the current is flowing toward you, the field lines (the direction of the force on a north magnetic pole) go in a counterclockwise direction, as shown in diagram **(a)** on the opposite page. (Throughout this book, the direction of an electric current is defined as being the direction of flow of a positive charge. However, scientists now know that it is a flow of electrons in the opposite direction. Despite this, the historical anomaly remains as the conventional description of electricity.)

The field lines of a wire that is going in a counterclockwise direction look very different from the shape of the field of a bar magnet

MAGNETIC FIELDS OF ELECTRIC CURRENTS

The magnetic field lines around a current-carrying wire are circular. With current flowing downward into a surface, the field lines run clockwise when seen from above. The fields of nearby currents combine in the patterns shown here. When the currents in two parallel wires are in the same direction, the wires attract each other. When they are in opposite directions, the wires repel each other.

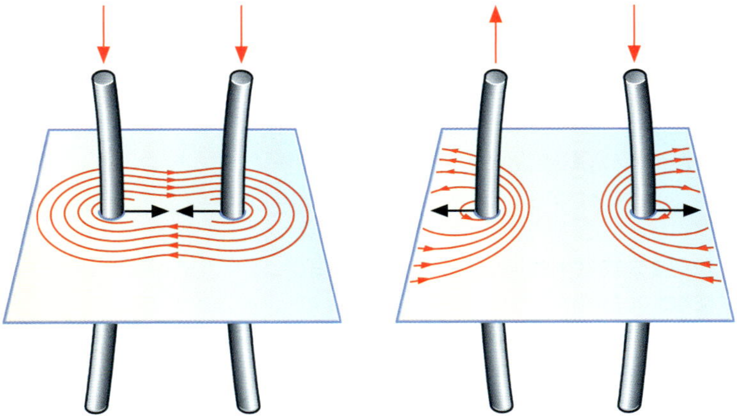

Metal detectors at airports are designed to detect concealed weapons and explosive devices in baggage or clothing. When a person walks through the detector, it responds to the magnetic effects of metal objects.

or of Earth. But if the wire is coiled into a single loop, the shape of the field begins to resemble that of a bar magnet (see **b**, at right). And if the wire is looped into many turns to form a cylindrical coil, called a solenoid (see **c** and **d**, at right), the fields of all the loops combine to make a stronger field, and the field lines strongly resemble those of a bar magnet.

Attraction and Repulsion

The field of a magnetic material is created by tiny electric currents in the atoms of metal that form tiny magnets. Together, they form one big magnet. The magnetic fields of electric currents interact with each other just like the magnetic fields of magnets. The result is that electric currents attract and repel one another. Currents flowing in the same direction in two wires attract each other. Currents flowing in opposite directions repel each other. Solenoids, like bar magnets, each have a north-seeking and a south-seeking end, or pole. Unlike poles attract, while like poles repel.

CURRENT-CARRYING COILS

The field of a current **(a)** becomes increasingly like that of a bar magnet as the wire is bent into more and more coils **(b**, **c**, and **d)**.

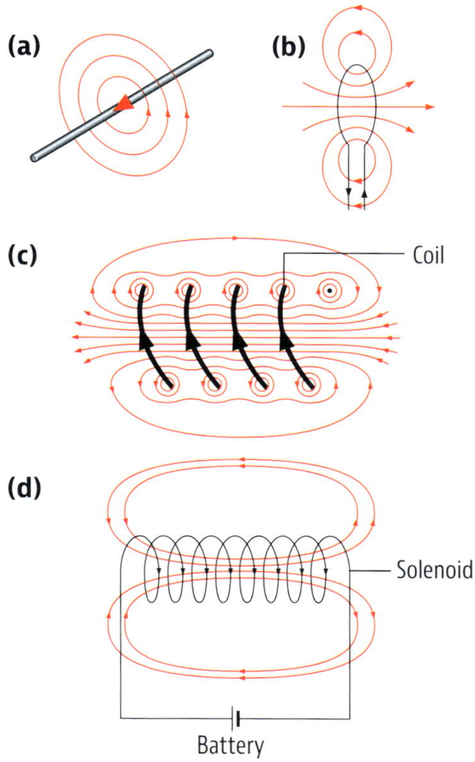

SCIENCE WORDS

Electromagnetism: The interlinked phenomena of electricity and magnetism. Every electric current generates a magnetic field, while changes in a magnetic field can cause a current to flow.

Solenoid: A current-carrying coil of wire. When a current flows through a solenoid, a magnetic field is developed in it. Often there is a moving iron core in the coil that moves when it is attracted by the magnetic field.

ELECTROMAGNETS

Electricity can be used to generate powerful magnetic fields that can be turned on and off at the flick of a switch. Electromagnets are valuable in industry, and the principle behind them is applied in countless other devices, many of which are used in domestic homes.

Coiling a current-carrying wire into a solenoid makes it into an electromagnet. This is a type of magnet that can be switched on and off. The strength of its magnetism depends on the strength of the current passing through it. But the amount of current is limited by the heating effect that it causes.

THE PRINCIPLE OF THE ELECTROMAGNET

A core of iron concentrates the field of a current-carrying solenoid, but it loses almost all of its magnetization when the current is switched off.

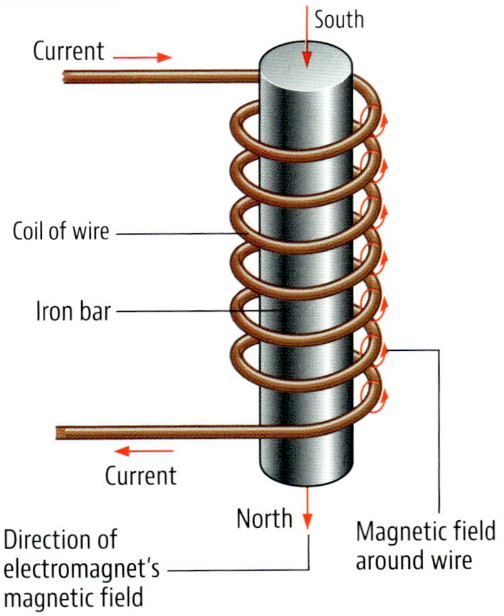

An electromagnet can be made more powerful by wrapping the wire around a central core of iron. It becomes strongly magnetized by the magnetic field of the solenoid, which greatly strengthens the total field of the magnet. As the strength of the current is increased, the magnetization of the iron increases until it reaches a maximum—it becomes saturated (see pages 4–5). Increasing the current further increases the field only by the amount contributed by the solenoid's own direct field. Whatever material is used in the electromagnet's core, it needs to be "soft," which means that it is able to lose magnetism as readily as it gains it.

The current required to power an electromagnet can come from any of several different sources. In portable devices such as electric razors, batteries do the job. On motor vehicles, ships, and airplanes the vehicle's own engine or engines generate an electrical supply that powers all the systems, including those that incorporate electromagnets. In homes, offices, and factories the electricity supply is the source of current.

The Compact Muon Solenoid, part of the Large Hadron Collider experiment at CERN in Geneva, incorporates a massive electromagnet 42 ft (13 m) long and almost 20 ft (6 m) wide in diameter. This solenoid's refrigerated coils are made from a niobium-titanium alloy.

In some of the first experiments to study electromagnets the U.S. physicist Joseph Henry (1797–1878) used a primitive electric cell, or battery. It consisted of plates of two different kinds of metal, such as copper and zinc, dipping into acid. Using hundreds of turns of insulated wire, Henry made electromagnets that could lift weights of over a ton.

Peril on the Sea

One of the strangest applications of the electromagnet is in wartime. Magnetic mines are tethered to the seafloor and float just beneath the surface of the water. When a ship passes nearby, the metal of the ship is detected by a sensitive magnetic needle in the mine. The swinging needle acts as the moving part of an electric switch. If a nearby

TRY THIS

Testing an Electromagnet

This project will test a simple electromagnet to see if it can be made stronger.

(Warning: do not leave any of the electromagnets connected to the battery for too long because the wires and the nail will become very hot.)

What to do

Wind a 4-ft (1.3-m) length of insulated wire around a steel nail of about 4 to 6 in (10 to 15 cm) in length, leaving about 1 ft (30 cm) of wire free at each end of the nail. Connect the free ends of the wire to the terminals of a 9-volt battery. To test the strength of this electromagnet, see how many paper clips the nail will pick up.

Now wind a piece of insulated wire twice as long around another nail, again leaving about 1 ft (30 cm) of wire free at each end. Again, connect the wires to a battery and see how many paper clips the nail will pick up. Did this nail pick up more clips or fewer clips than the first electromagnet?

Finally, connect each electromagnet to two batteries (below). Which electromagnet picked up the most paper clips? Were the electromagnets with two batteries more powerful than those with only one? You should find that the greater the number of turns of wire, the more powerful the electromagnet. Also, the higher the voltage, the more powerful the electromagnet.

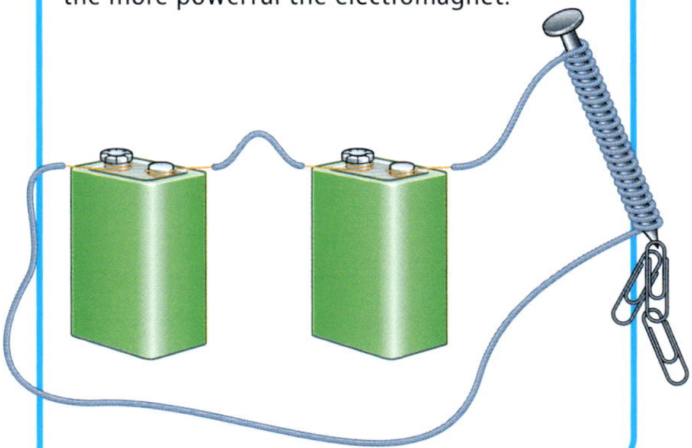

JOSEPH HENRY

The American physicist Joseph Henry discovered the principle of electromagnetic induction—that changing electric currents in one circuit can cause currents to be generated in another circuit. However, Henry did not get the credit for the discovery because the British physicist Michael Faraday (1791–1867) announced the results of his own studies before him. Henry developed improved electromagnets and built one of the first electric motors. He also invented one of the first successful electric telegraphs, allowing messages to be sent over long distances without them growing fainter. In 1846, Henry became first director of the newly formed Smithsonian Institution in Washington, D.C., and served in that position until his death in 1878. He developed the study of meteorology at the Smithsonian and pioneered weather forecasting.

PRINCIPLE OF HENRY'S ELECTROMAGNET

Coils magnetize the two legs of the horseshoe magnet. The magnetism in each leg makes a magnetic pole at each end of a piece of iron placed across the horseshoe's poles.

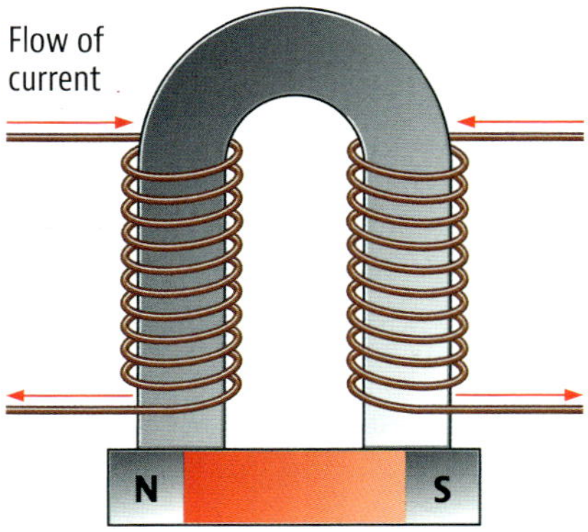

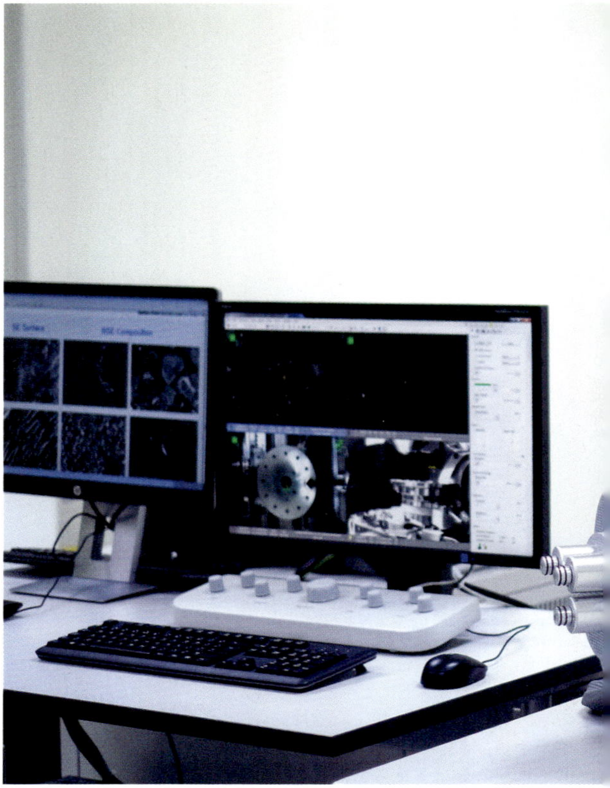

ship makes the needle swing, the switch is closed. Current from a battery flows through an electric detonator, exploding the mine and damaging or sinking the ship. The mine is triggered because the ship is a giant magnet with a magnetic field induced by long exposure to the Earth's magnetic field. Degaussing belts—which are cables that encircle the ship's hull and carry electric current—generate a magnetic field that cancels out the ship's own magnetism and protects it from detection by mines.

Beam Benders

Electric currents are not confined to those that flow in wires. They also flow through free space. In old-style television sets, a beam of electrons traveled from a heated cathode at the rear of the set and swept across the screen to illuminate tiny phosphor dots on

A scanning electron microscope (SEM). Magnification is up to 250 times better than the best optical microscopes.

the screen, so forming a picture. Beams of electrons are used in electron microscopes, which are instruments that can see details far smaller than can be seen with optical microscopes using ordinary light.

One type of electron microscope, the transmission electron microscope, or TEM, can see details the size of individual atoms. The electrons are accelerated by hundreds of thousands of volts. (The electron beam current may be only a billionth of an amp, compared with about half an amp in the current going through a typical desk lamp bulb.) The electron beam is shaped, focused, and guided by electromagnets consisting of current-carrying coils through which the beam passes on the way to its target.

ELECTRON MICROSCOPE

This uses a beam of electrons in place of a beam of light. Electrons, like all subatomic particles, have a wavelength that is much shorter than that of light, so making it possible to form extremely detailed images. The electron beam traveling from the top of the instrument is focused by strong "magnetic lenses," which are current-carrying coils.

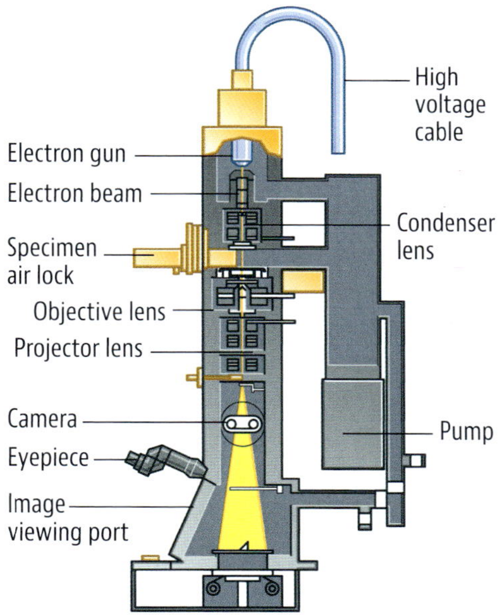

SCIENCE WORDS

Horseshoe magnet: A permanent magnet that is bent into a horseshoe shape so that the two poles are close together, and their strengths combine.

Magnetic lens: An arrangement of electromagnets that can focus a beam of electrons or other charged particles, as in an electron microscope.

Magnetization: The process of making an object or material into a magnet.

ELECTROMAGNETIC DEVICES

Today is a push-button world—it seems normal that the gentle pressure of a finger is enough to lift mighty weights or to carry words and images all around the world. However, this was only made possible by putting electromagnetism to work.

The discovery of electromagnetism by scientists in the early 19th century led to a steady stream of inventions that depended on the interaction of electric currents and magnetic fields. This stream grew into a flood, and today electromagnetic devices are part of daily life. The American physicist Joseph Henry (1797–1878) made one of the earliest such innovations in 1831, when he invented the first electric bell. A bell was struck repeatedly by a clapper held away from the bell by a spring if no electric current flowed. Pressing a button allowed the electric current to flow in the circuit through a solenoid. The magnetic field generated by the solenoid attracted the metal arm of the clapper, which struck the bell. But the movement of the clapper also broke the circuit so that the current stopped and the magnetic field disappeared. The clapper, no longer attracted, jumped back and closed the circuit. As long as the button was pressed, this cycle repeated itself.

The electric bell was much more convenient than the pull-cord bell in use in the 1800s. Electric wires could run from room to room more easily than pull-cords, and could sound more than one bell if needed. Bells could be very loud and have a variety of tones. Variants, such as chimes, were also provided.

The Telegraph

An early use of electromagnetism was in the electric telegraph, which sent messages by wire over long distances in the form of long and short pulses of electric current. These pulses moved a recording pen at the receiving end. Short electric pulses were registered as "dots," and long pulses as

SAMUEL MORSE

Samuel Finley Breese Morse was born in Charlestown, Massachusetts, in 1791. After graduating from Yale in 1810, he studied art in England before returning home to become a portrait painter. In the 1830s, his interest turned to science; he was inspired to invent the electric telegraph after meeting Joseph Henry, who showed him that the arrival of an electric current along a wire could be detected by making it work an electromagnet. When the current arrived, the magnet pulled over an iron armature (current-carrying wire coils in a motor or magnetic device that produce force) with an audible "click." In 1843, Morse built a 37-mile (60-km) long telegraph line from Washington, D.C., to Baltimore. In 1844, he sent the world's first message ("What hath God wrought!") via telegraph. Essential to this technology's success was Morse code, a system of short and long electrical pulses that encode the letters of the alphabet. The code was devised by Morse and his assistant Alfred Vail (1807–1859). Samuel Morse died in New York City in 1872.

An electric telegraph key (left) and receiver (far left). As the pulses were registered by the receiver, it marked the dots and dashes of the message on a paper tape.

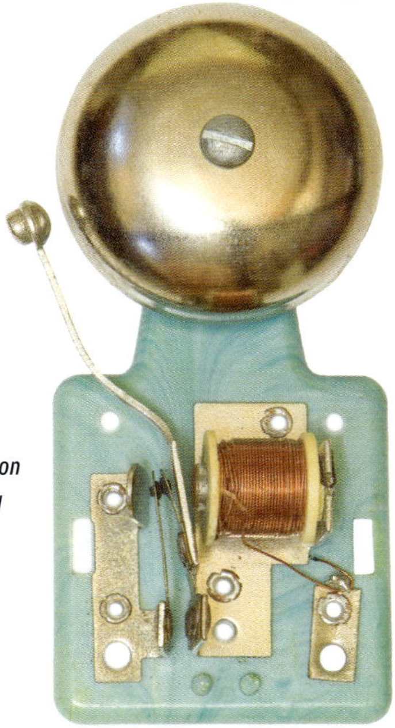

The electric bell vibrates by the action of a hammer pulled one way by an electromagnet and the other by a mechanical spring.

"dashes." The combinations of dots and dashes stood for letters of the alphabet according to a code that was devised by the U.S. inventor Samuel Morse (1791–1872).

To send signals over long distances, it was necessary to overcome the weakening of signals as they passed over telegraph wires many miles long. Electromagnetism was put to use again to solve this problem with the invention of the "relay." The signals in each stretch of telegraph wire, roughly 20 miles (32 km) long, were used to open and close switches in the next stretch of the wire, which had its own electricity supply. In this way, signals could be sent for unlimited distances without any loss of strength.

This principle is still important today. A relay is a device that enables the current in one circuit to control the current in another. The second current may be large and dangerous, while the first one is small and safe to operate. When a truck driver brings an engine to life with a powerful current from the battery, it is controlled by a weak current activated by the turn of the ignition key.

DOOR CHIMES

Pressing the doorbell sends a pulse of current through the solenoid, making the iron rod jerk across sharply and hit one of the chimes. Instantly a spring (not shown) pulls it back again, striking the second chime.

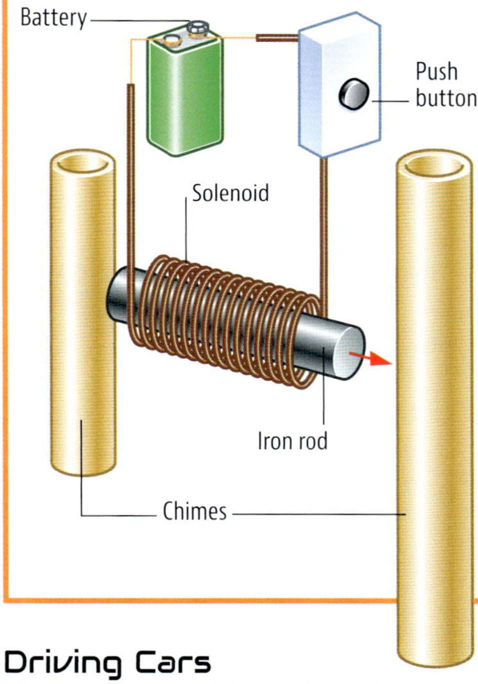

Driving Cars

Magnets are essential in today's electric cars, which are powered by electricity rather than gas and have large batteries that require magnetism to produce power. Vehicles with internal-combustion engines are also packed with devices in which magnetism plays a key role. Many of these devices are small motors (see pages 34 to 37). There is also a variety of magnetic sensors. One detects the position of the throttle in cars with cruise control, keeping a constant throttle setting without the need for foot pressure. Another sensor detects the position of the steering wheel in cars with power steering, so that the amount of power that is delivered to help turn the wheel can be controlled precisely.

STARTER MOTOR

Gas-powered cars have an electric starter motor, which is turned by a large electric current from the car's battery. This current is turned on by the starter relay, in which a low current from the battery activates a solenoid when the driver turns the key in the ignition switch.

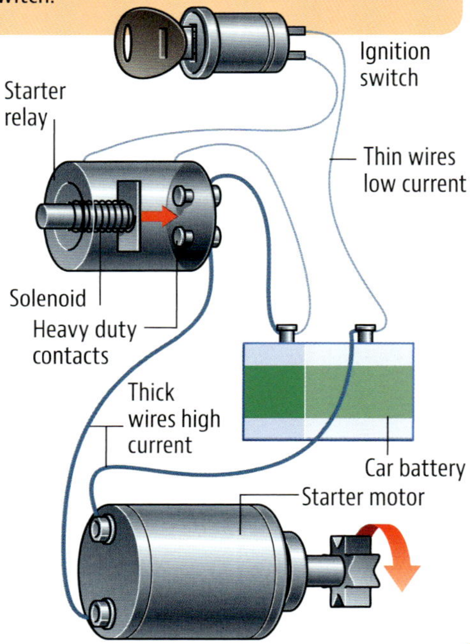

Electromagnetism and Sound

Magnetism is vital in sound reproduction. Relatively weak electric currents, varying in strength many thousands of times per second, drive loudspeakers in a radio, cell phone, or TV set. The pattern of the current—the audio signal—is a copy of the pattern of loudness of the voices, music, or other sounds that the current might be representing.

This varying current is passed through a solenoid called the voice coil, attached to the center of a paper or plastic cone. The coil is in the field of a strong permanent magnet that may take the form of, say, a ring surrounding the coil. The varying current gives rise to a constantly fluctuating magnetic field generated by the coil, and the coil is pulled by the permanent magnet with a strength that depends on the current. The coil vibrates, along with the cone attached to it. The vibrating cone disturbs the air, setting up sound waves that are exact copies of the original ones.

A microphone originally produces the electric current that carries the audio signal. In the most important types of microphone, once again it is electromagnetism that plays a key role. The vibrations of sounds that strike the microphone cause a light plastic diaphragm to vibrate. Attached to the diaphragm is a metal ribbon or coil in the field of a permanent magnet. As the metal vibrates in the field, the effect is the same as if the magnet were being moved while the metal remained still: a voltage is induced in the metal. This creates the small audio current.

The first sound-recording machines had speaking-tubes for microphones and large

Speakers such as these, which have built-in amplifiers, contain a strong permanent magnet and a coil acting as an electromagnet. The interaction between them vibrates a cone in the speaker and generates sound.

horns to amplify the faint sounds that were produced by the needle as it ran along the groove of a recording disk or cylinder. The electronic reproduction of sound proved more accurate, allowing for a greater range of frequencies and a greater volume when this was required.

Electromagnetism was used in a different way to revolutionize music. Magnetic detectors called pickups were added to the steel-stringed guitar. This turned the vibrations of the strings into electrical signals and the electric guitar was born. Electric guitars could be amplified (made stronger) or the sounds altered in various ways and then fed into a loudspeaker. This new power enabled the guitar to play a prominent role in jazz, blues, rock, and pop music.

SCIENCE WORDS

Current: A flow of electric charge (electrons). Current may be supplied by the domestic electricity supply, generators, or batteries.

Motor: A machine that converts energy, usually electrical energy, into motion.

Relay: A device that is activated by changes in an electric current in a circuit, causing it to switch a second electric circuit on or off. Relays often incorporate electromagnets.

THE RELAY

In central heating programmers, a small current (electronic circuit) flips on the power that controls a large current (gas boiler). This is a relay device. Below, a switch in a low-current circuit controls a motor that draws a high current. Closing the switch in the lower circuit allows a low current to pass, so the solenoid generates a magnetic field that attracts the pivoted armature, allowing a strong current to flow in the second circuit to operate the motor.

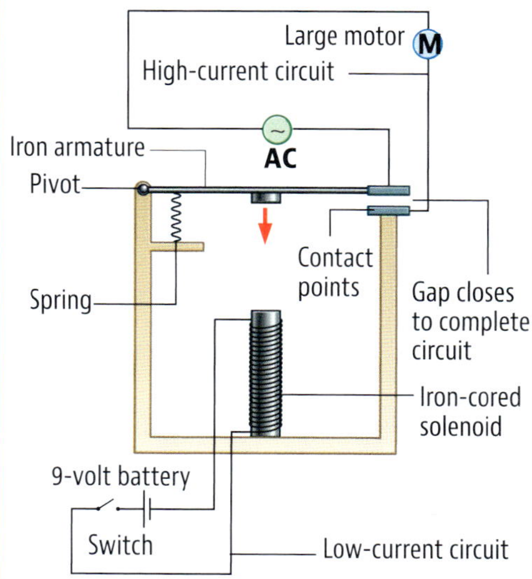

MOTOR EFFECT

Magnetic fields that are produced by permanent magnets or electric currents can move current-carrying wires. Early researchers studied this effect and designed the first electric motors. Their laboratory toys became the workhorses of modern industry.

After Ørsted's ground-breaking discovery of 1819, the British physicist Michael Faraday (1791–1867) carried out the most thorough experimental investigation of the interactions between electricity and magnetism. The French physicist André Ampère (1775–1836) had found that not only did current-carrying wires produce magnetic fields, but that they also experienced forces when they were placed in magnetic fields, making them seem even more similar to magnets. Faraday announced his "left-hand motor rule" as a reminder of how these forces acted. Imagine that with the palm facing you, the thumb, first finger, and second finger of your left hand are extended at right angles to one another:

- if the first finger represents the direction of the field
- and the second finger represents the direction of the current
- then the thumb represents the direction of movement of the current-carrying wire.

(As always in this book, the direction of the current is taken to be the direction

MOTOR EFFECT DEMONSTRATION

When current flows through the copper rod that is dipping into the mercury, the rod revolves around the magnet.

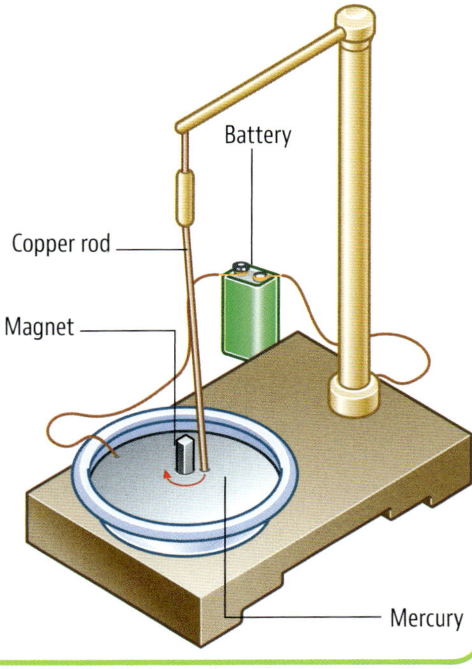

Michael Faraday (left) in his laboratory with John Daniell, inventor of a battery that bears his name *(see page 59).*

SCIENCE WORDS

Coil: A spiral of wire through which current flows. The magnetic fields of the current in the different turns of the coil add together to make a large magnetic field. The coil is then an electromagnet, with a field resembling that of a bar magnet.
Left-hand motor rule: When the directions of a magnetic field and electric current are represented by the first and second fingers of the left hand, the direction of motion of the current-carrying conductor is indicated by the thumb when all three digits are extended at right angles to one another.
Torque: Any force or system of forces that cause rotation or twisting.

of flow of positive charge, the opposite of the direction of flow of the negatively charged electrons.)

A Primitive Motor

Faraday set up a striking demonstration of the effect of the interaction between current and magnetism. He let a current flow through a freely suspended copper rod dipping into a pool of mercury (a good conductor). When a bar magnet was placed upright near the rod, the rod started to revolve around the magnet and kept revolving for as long as the current flowed. Because the current was vertical and the magnetic field lines went to and from the magnet, there was a force on the rod at right angles to both of them—that is, around the circumference of a circle.

If a current passes through a coil in a magnetic field, the two sides of the coil experience forces in opposite directions because the current flows in opposite directions on the two sides. This creates a torque, or twisting force, on the coil except when it is at right angles to the field lines, when the forces stretch or squeeze the coil, but not turn it. The more turns of the coil, the stronger the current, the magnetic field, and the force on it. This was to be the basis of the powerful electric motors of today.

MAGNETIC FORCE ON A CURRENT

Faraday showed that a current-carrying wire experiences an upward force when the current and magnetic field are in the directions shown here.

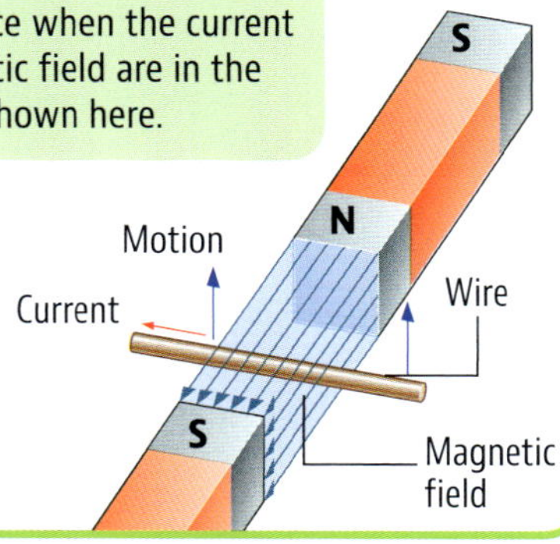

MAGNETIC FORCE ON A COIL

A current-carrying coil in a magnetic field tends to rotate until it is at right angles to the field lines. The interaction between the magnetic field and the current causes the coil to turn **(a)**. When the coil is at right angles to the field, there is no turning force **(b).**

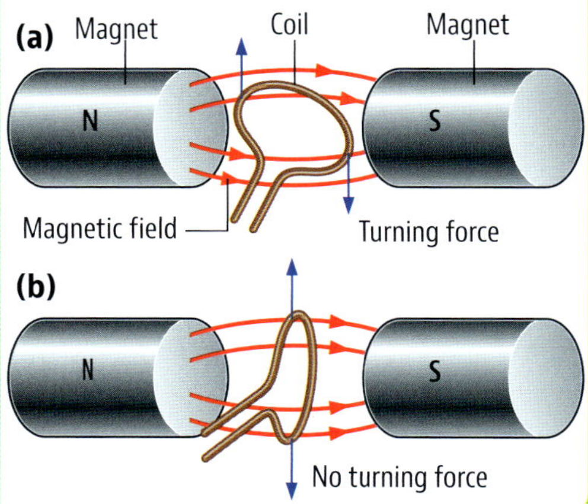

ELECTRIC MOTORS

Small electric induction motors. The wound copper coils are clearly visible on either side of the spindles. When these are supplied with power, they create a rotating magnetic field pattern that in turn gives rise to current in the rotor conductors, causing the rotors to turn.

Modern life would be very different without electric motors. There are many types of motor and each is designed to carry out a particular task, from small motors in electric clocks to large winding motors for the elevators in a skyscraper. All motors are the result of ingenious engineering that exploits the close connection between electricity and magnetism.

The simple rotating coil shown on the previous page has to be made considerably more complicated to turn it into an electric motor. Yet the basic principle remains the same in all motors: an electric current moving through a magnetic field experiences a force at right angles to itself and to the field. The force can be turned into movement, and the movement can be made to do useful work such as moving a vehicle, raising a heavy weight, or opening a door.

The DC Motor

Direct current (DC) is a steady current that does not vary with time. In the simplest sort of electric motor, direct current is passed through a coil of many turns that is free to rotate in the field of a permanent magnet. In the upper illustration on the opposite page, current flowing away from you experiences an upward force. With the coil in the position shown, the torque is therefore clockwise. When the plane of the coil is parallel to the field, like this, the torque (twisting force) on the coil is at its maximum.

When the coil has turned through 90 degrees from this position, there is no torque on it, but its momentum carries it past this point. If the flow of current were not altered, the torque on the coil would now be counterclockwise, slowing down the rotation. To stop this happening, the current is reversed when the coil is at 90 degrees to the field. The current is supplied through a commutator—a cylinder split into two parts, each part touching one brush, or contact. As the coil turns to the 90-degree position, each half of the commutator comes into contact with the other brush, reversing the direction of flow into the coil. The coil gets an increasing twist in the same direction as it turns through the next 90 degrees. When it has turned to the point where it is once again at right angles to the field, the current is again reversed.

The AC Motor

Alternating current (AC) is a current that changes many times each second. An AC motor can in principle be simpler than a DC motor. Because alternating current reverses itself periodically (the electricity supply goes through 60 cycles per second in the United States), there is no need for a commutator when the coil revolves at this rate. Such a motor adjusts its rate automatically to the frequency of the supplied current.

SCIENCE WORDS

AC motor: An electric motor that is operated by alternating current.
Alternating current (AC): Electric current that flows first in one direction, then in the other, alternating many times each second. AC is used for domestic electricity supply and many other electrical applications.
DC motor: An electric motor that is operated by direct current.
Direct current (DC): Electric current that flows in one direction all the time, though it may vary in strength.

DC MOTOR

The magnetic field forces the current-carrying coil to rotate. The split-ring commutator is fixed to the coil and reverses the current every half turn.

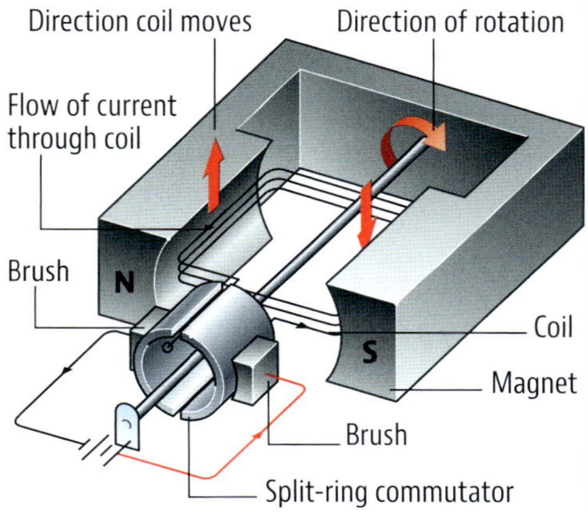

AC MOTOR

The direction of the current reverses continually, ensuring that the coil always experiences a turning force in the same direction.

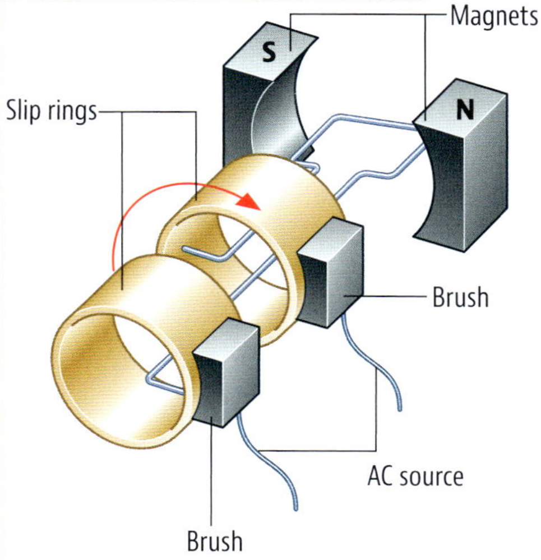

INDUCTION MOTOR

AC current flowing in the coils of the stator (the outer assembly) generates a rotating magnetic field that drags the rotor around.

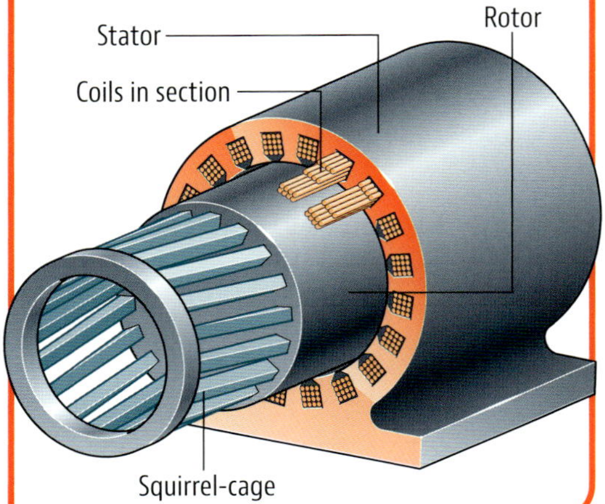

However, all practical motors have to be more complicated than the previous examples. The magnetic field is likely to be supplied by electromagnets rather than by a permanent magnet. The power supply to the motor sends some current through coils called field magnets around the outside of the motor in an assembly called the stator. There can be many rotating coils, making up an assembly called the rotor. It is supplied with current through a commutator divided into a corresponding number of segments, so that at any moment current flows into the coil that is in the plane of the field and so produces the maximum torque.

Induction Motors

The most common type of AC motor is the induction motor, which works on a different principle from the one described above. In an induction motor, AC current is passed through field coils in a stator, one coil after another, so that viewed from one end, the direction of the magnetic field rotates. Housed within the stator is a structure called a squirrel-cage, which consists of bars of iron or other soft magnetic metal. The revolving field in the stator induces a magnetic field in the squirrel-

LINEAR INDUCTION MOTOR

If an induction motor **(a)** were imagined to be "unrolled" **(b)**, it would form a linear induction motor **(c)**. Traveling magnetic fields in the track pull the metal plate along.

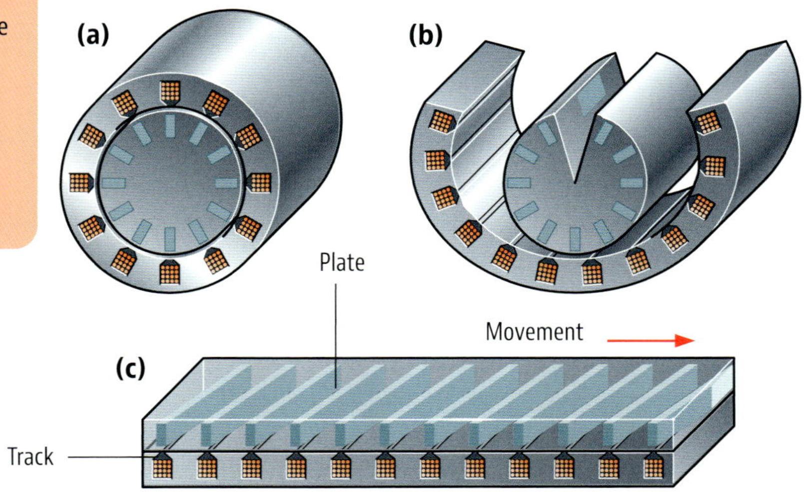

PRACTICAL DC MOTOR

The magnetic field is provided not by a permanent magnet, but by current passing through field magnet windings. The commutator has many segments, which correspond to the number of field magnet coils contained in the armature.

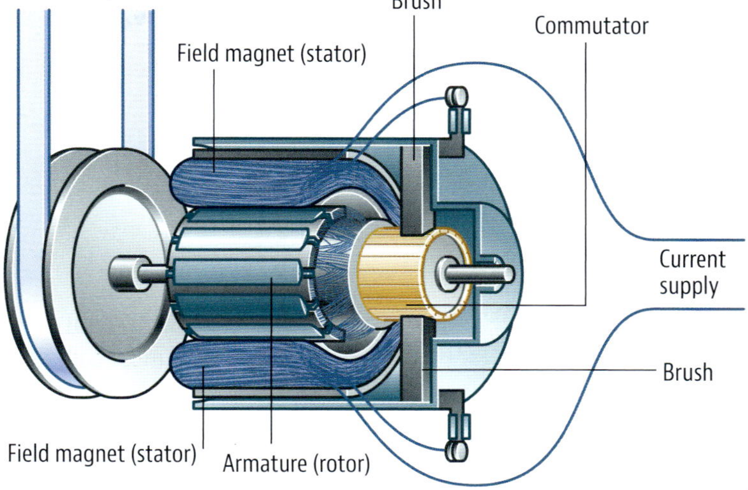

cage (hence the name "induction motor"). The squirrel-cage moves around with the field. A similar principle is used in motors that move objects in straight lines.

In the linear induction motor, field coils are arranged along a track. A pulse of current passes through successive coils in turn, sending a wave of magnetism along the track. A flat plate of metal laid on or alongside the track has magnetism induced in it and moves along with the wave. These types of motors are used for sliding doors and in some other machines, such as the shuttles of looms, where straight-line motion is required.

Motors Everywhere

In industrialized countries, electric motors are to be found everywhere. Most domestic appliances—including dishwashers, washing machines, food-blenders, juicers, drills, and vacuum cleaners—rely on electric motors. Even a desktop computer has motors in it to power hard disk drives and cooling fans. In factories, a host of machines are driven by electric motors, and they power transportation systems, such as trains and trams. In cars, small electric motors drive windshield wipers and washers, electric windows, sunroofs, centrally controlled door locks—and even adjustable rear-view mirrors, seats, and the extending radio antenna. And every one of these electric motors relies on electromagnetism.

SCIENCE WORDS

Induction motor: An electric motor in which a moving magnetic field induces (generates) current in another component, which moves to follow the field. The moving part may revolve or move in a straight line. See also Linear induction motor.

Linear induction motor: An induction motor that produces straight-line motion, as, for example, in making a sliding door open and close.

Stator: The stationary part of a motor or generator.

USING ELECTRIC MOTORS

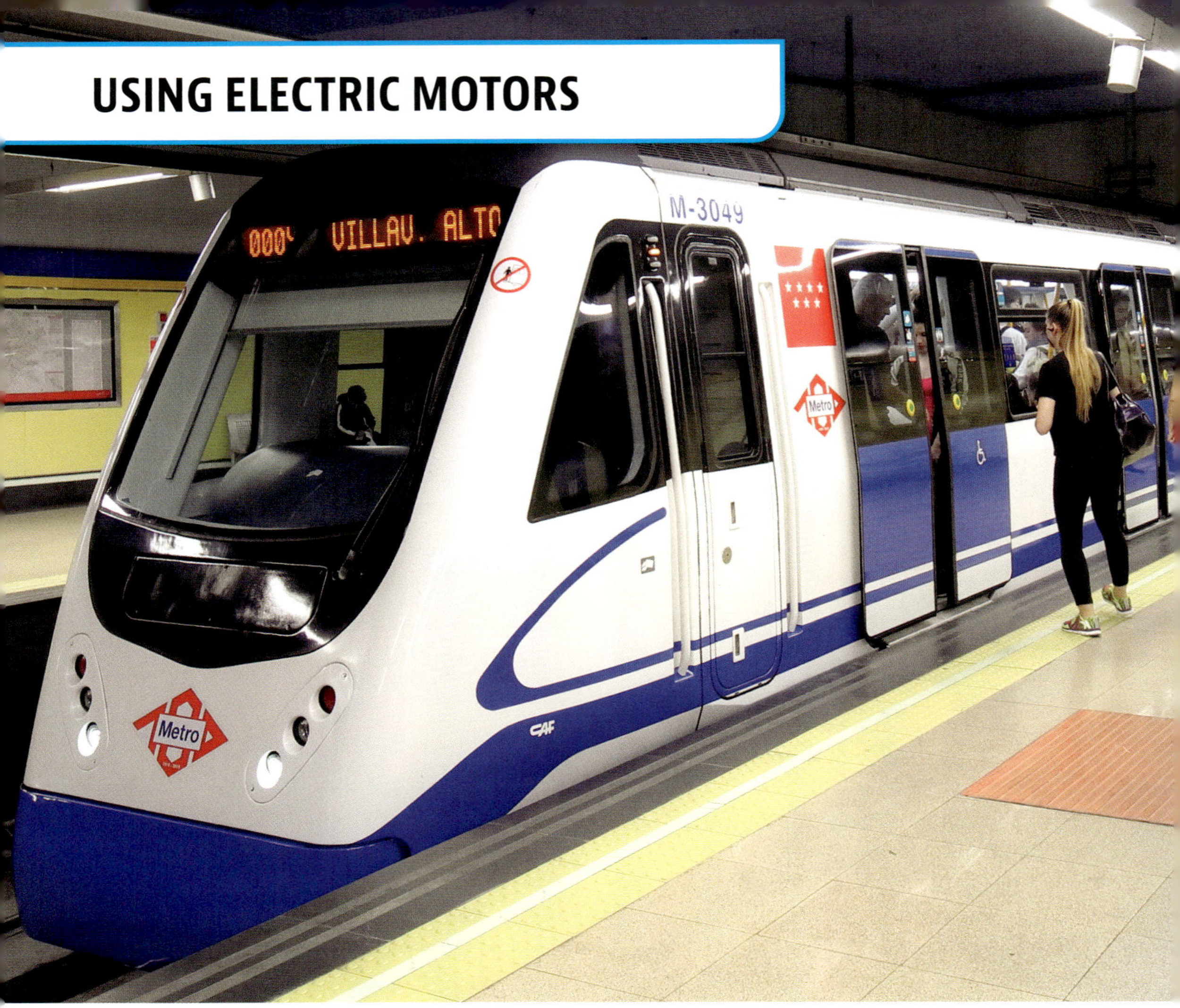

Electricity is a versatile form of energy that can be harnessed to do all kinds of things. An electric motor is a machine that converts electrical energy into mechanical energy. Motors are used everywhere; in the home, the garden, and in industry.

Industrialization began in the 18th century, using water power to drive machinery in mills and other types of factories. Water power was followed by steam power. However, it was the development of the electric motor in the 19th century that carried industrialization into a new phase of productivity and paved the way for the development of the modern world. By 1850, experimental electric motors were being used to power boats, ships, and railroad locomotives.

Today, electric motors do a diverse range of jobs. Some are designed to move heavy weights over large distances, such as the motors that raise passenger elevators in tall office buildings. Some have to provide smooth movement, such as the motors that operate sliding doors; while with others, such as the powerful motors used in garbage-disposal units, the strength of the electricity produced is all important. With electric motors used in industry, the size may be limited by the cost

The first subway trains in London, England, were powered by steam, which created major ventilation problems. From the late 19th century, electric traction was used in underground systems worldwide, as in this train on the Madrid Metro in Spain.

SCIENCE WORDS

Linear induction motor: An induction motor that produces straight-line motion as, for example, in making a sliding door open and close.

Pole: One of the two regions of a magnet where the strength of the field is greatest. Lines of force diverge (radiate out) from one pole and converge on the other.

Rotor: In electromagnetic technology the rotating coil in a motor or generator. See also Stator.

Stator: The stationary part of a motor or generator. See also Rotor.

of large motors. Conversely, some electric motors are extremely small, like those that are used inside watches, electric toothbrushes, or can openers.

Ups and Downs

In 1857, a steam-operated passenger elevator went into service in a New York City department store. However, although steam-operated elevators were successful, electrically powered elevators replaced them in the late 1880s. These were cleaner, quieter, and more powerful, and different types of motor were developed to allow the cars to operate at a range of speeds without needing complicated gearing. The motors driving the fastest modern elevators can move them at speeds of up to 28 mph (45 km/h).

Tall, modern buildings need so many elevators that engineers are designing them to occupy smaller amounts of space. One way is to remove the motors from the winding gear at the top of the elevator shaft and replace them with flat linear induction motors that are built into the elevator car.

Stopping and Starting

There are countless variations in the design of electric motors, depending on the different circumstances in which they are to be used. There may be special adjustments to help them start slowly or prevent them from starting too quickly and violently.

Electrical current going to a motor's stator (pages 34–37) flows through the field coils, creating a magnetic field directed across the stator. In the simplest type of motor, called a two-pole motor, this field constantly reverses through 180 degrees because the supply is AC, and so reverses continually. The rotor has its own magnetic field, which is produced either by a permanent magnet or by current. It swings around repeatedly, "trying" to bring its own north pole close to the south pole of the stator field, but never succeeding because the stator field is constantly switching.

This kind of motor can be slow to get started. If, when the motor is switched on, the rotor field happens to be lined up with the stator field, then when the stator field flips, the rotor does not get a twist to one side or the other. Once the rotor is moving there is no problem—its own momentum carries it past this neutral position. Also, it can rotate equally well in either direction, so it has to be given an initial nudge in the right direction. Such a motor comes with an additional field coil and circuits that create an extra field when the device is started up; once the motor is up to speed, they are disconnected automatically.

Often a motor has more than two poles—that is, more than one pair of field coils. As explained earlier (pages 34–37), current is sent to these pairs in succession to create a magnetic field that rotates, as seen from one end of the motor. This kind of motor generally has no startup problem.

In some kinds of motor there is a different kind of problem on starting. The motor is liable to start moving violently if full voltage is applied to it on switch-on. This would not be acceptable in streetcars and elevators, say, which must not jolt violently. Additional circuits are built into motors for such applications, which limit the field that is developed in the motor's rotor immediately after switch-on and are disconnected later.

Keeping Steady

There are many applications in which it is important for the motor to turn at a steady speed. One is any kind of sound recording and playback, whether in a traditional phonograph

FAN MOTOR

An induction motor usually drives a fan, such as the one in a blow dryer. Alternating current is supplied to field coils that produce a rotating magnetic field. The rotating field turns an aluminum rotor on which the fan is mounted. There may be a speed control that alters the rate of rotation of the field and therefore of the fan.

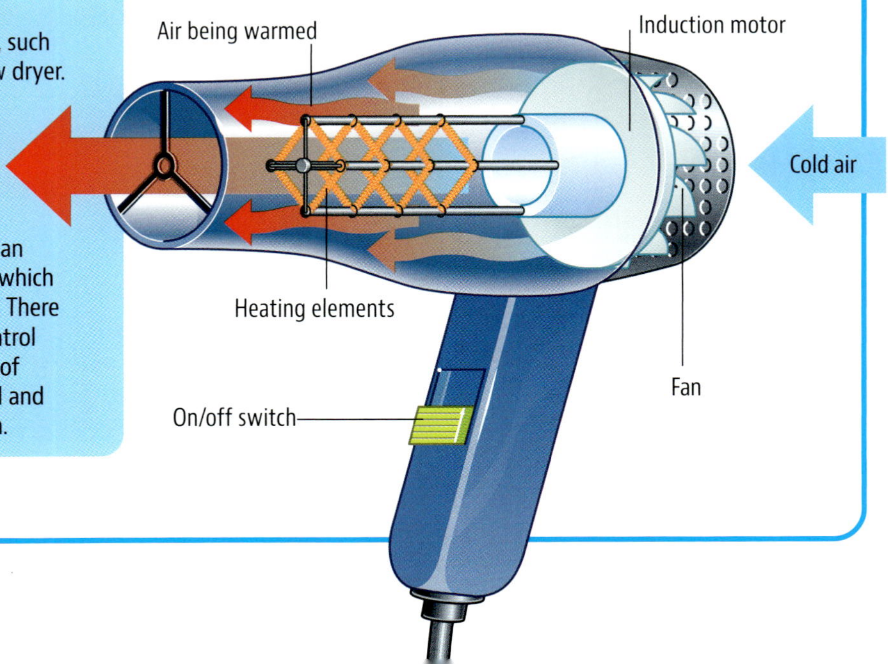

turntable, or a CD or DVD drive. Another is in clocks and watches (the analog sort, which have hands). The AC supply from the power company is an ideal "pacemaker" for a motor because it is delivered at a constant frequency (60 cycles a second in the United States). If it is supplied to the stator of a motor, it will generate a field rotating 60 times a second. If the rotor is equipped with a permanent magnet, or with an electromagnet supplied by a fixed-strength direct current, the rotor will follow the rotating field at the same rate. Such a motor is called "synchronous."

Motors in watches and other portable devices have to be powerful yet compact. When a cell phone is on "silent," a tiny electric motor makes the phone vibrate. The head of the motor is slightly bigger on one side than on the other, so it wobbles and shakes as it rotates (a bit like an unevenly loaded washing machine).

Electric drills usually have a "universal" motor that runs on both AC and DC current, with two speed settings. The drive unit in a DVD player (below) is controlled to run between 200 and 500 rpm (revolutions per minute), depending on the track being read.

ELECTRICITY FROM MACHINES

Despite the increasing shift to new forms of renewable energy, fossil fuels are still used to make electricity. Electricity generating power plants use monster machines and consume huge amounts of fuel to meet the growing demands for energy by expanding industries worldwide.

In a power plant, turbines are spun by steam to generate electricity. The steam has to be produced by boiling purified water circulating in a closed system of pipes. Plants differ in the source of the heat they use to boil the water. Some burn coal; others, oil or gas; still others use heat from a nuclear reactor. A few use alternative forms of energy, such as sunlight or wave power, or heat from rocks deep beneath the ground. The steam used in a conventional or nuclear power plant is superheated to temperatures of about 1,000°F (550°C) or more. This steam is forced into the turbine, which consists of rotors, sets of rotating wheels equipped with fan blades, alternating with sets of fixed fan blades called stators. The pressure of the steam forces the turbine's shaft to spin. The temperature and pressure falls, but it is then led through lower-pressure stages of the turbine to extract as much energy as possible. The last of the steam's heat is removed as it is circulated through pipes that are cooled by water from some external source, such as a lake or river.

SCIENCE WORDS

Transformer: A device that increases or decreases the voltage of an alternating current. See voltage, below.
Turbine: A machine in which the energy of a moving fluid or air flow is converted into mechanical energy by causing a bladed rotor to rotate.
Voltage: (also called electromotive force, EMF.) The potential difference in electrical charge between two points in an electrical field. It is measured in volts. The greater the voltage, the higher the force moving two charges between the points.

Repeating Cycle

The condensed water is returned to be heated again and the cooling water is returned to its source, although its possible

Steam-driven turbines generate electricity. The main energy sources are coal, oil, and nuclear energy.

effect on the environment should be carefully monitored and minimized. Warm water encourages the growth of algae, which cuts down oxygen in the water used by fish. Water vapor is often seen rising from the cooling towers that are a feature of power plants.

A shaft runs from the steam turbine. The shaft spins at 3,600 rpm (in the United States) and is part of the generator. There is a strong magnetic field in the generator, which is produced by an independent supply of current flowing through electric coils called field windings. An armature, carrying electric coils, spins in this magnetic field. An electromotive force of about 25,000 volts is generated.

Because the shaft rotates at 3,600 rpm, the current alternates at 60 Hz (that is, 60 hertz, or 60 times per second). This means that during 1⁄120 second the current flowing in one direction increases and then decreases to zero; and then during the next 1⁄120 second the current builds up to a maximum in the reverse direction before decreasing to zero again. This is an alternating current (AC).

Stepping Up

The power is delivered from the generator at about 25,000 volts. It is sent to a transformer, which steps up the voltage to several hundred thousand volts. This is necessary for sending power over long distances. The electrical cables have resistance, and heat is generated in them, which is wasted. Having a very high voltage driving a low current can minimize the heat lost. The high-tension ("tension" here just means voltage) lines fan out from the plant in all directions to cities and towns in the area that it serves.

These voltages are extremely dangerous, and the wires must be carried on tall towers (pylons) well off the ground. They need extra support where they pass over highways. Ceramic insulators keep the current-carrying wires separated from the towers on which they are supported. In some places, high-tension cables are carried in pipes underground.

Stepping Down

Near an area where there are factories or other industrial plants, power cables lead to an unmanned substation. Here, transformers reduce the voltages. A variety of different voltages are produced to meet the needs of different users. The substation runs automatically. It is securely fenced off because it is extremely dangerous for people to wander around or play near it. Where power is to be delivered to homes for domestic use, the voltage is finally reduced to between 100 to 130 volts in North America and the Pacific, and 220 to 240 volts in other countries. The power is taken to an individual home along an overhead cable or under the street.

ELECTRICITY GENERATING PLANT

A power plant converts the stored energy of its fuel into steam, which a steam turbine and generator uses to produce a supply of electric current. A high proportion of the fuel's energy is inevitably lost as waste heat. Transformers boost the voltage for long-distance transmission. At substations, the voltage is stepped down to supply industry, and stepped down still further to supply homes and offices.

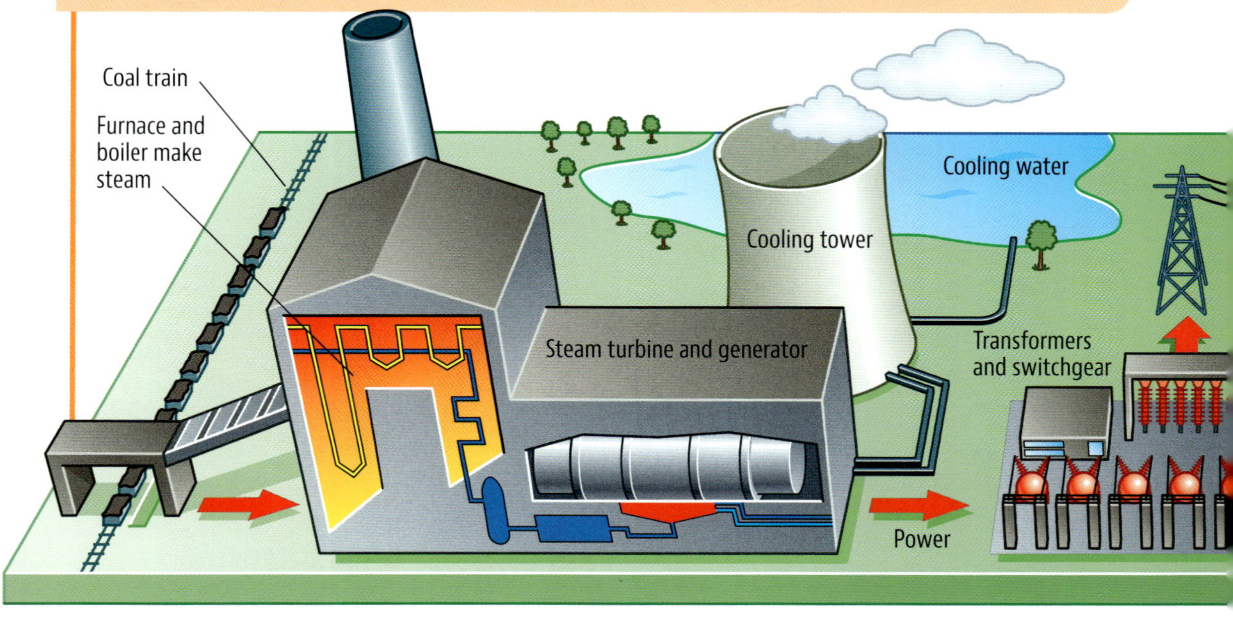

Transformers at substations or on high-voltage lines (left) step the voltage of the power supply down for use in homes.

SCIENCE WORDS

Alternating current (AC):
Electric current that flows first in one direction, then in the other, alternating many times each second. AC is used for domestic electricity supplies and for many other electrical applications.

Direct current (DC):
Electric current that flows in one direction all the time, although it may vary in strength.

Insulator:
A material that is a poor conductor of electric current. Examples of this are rubber, many plastics, and wood. ("Insulator" is also the name given to a poor conductor of heat.)

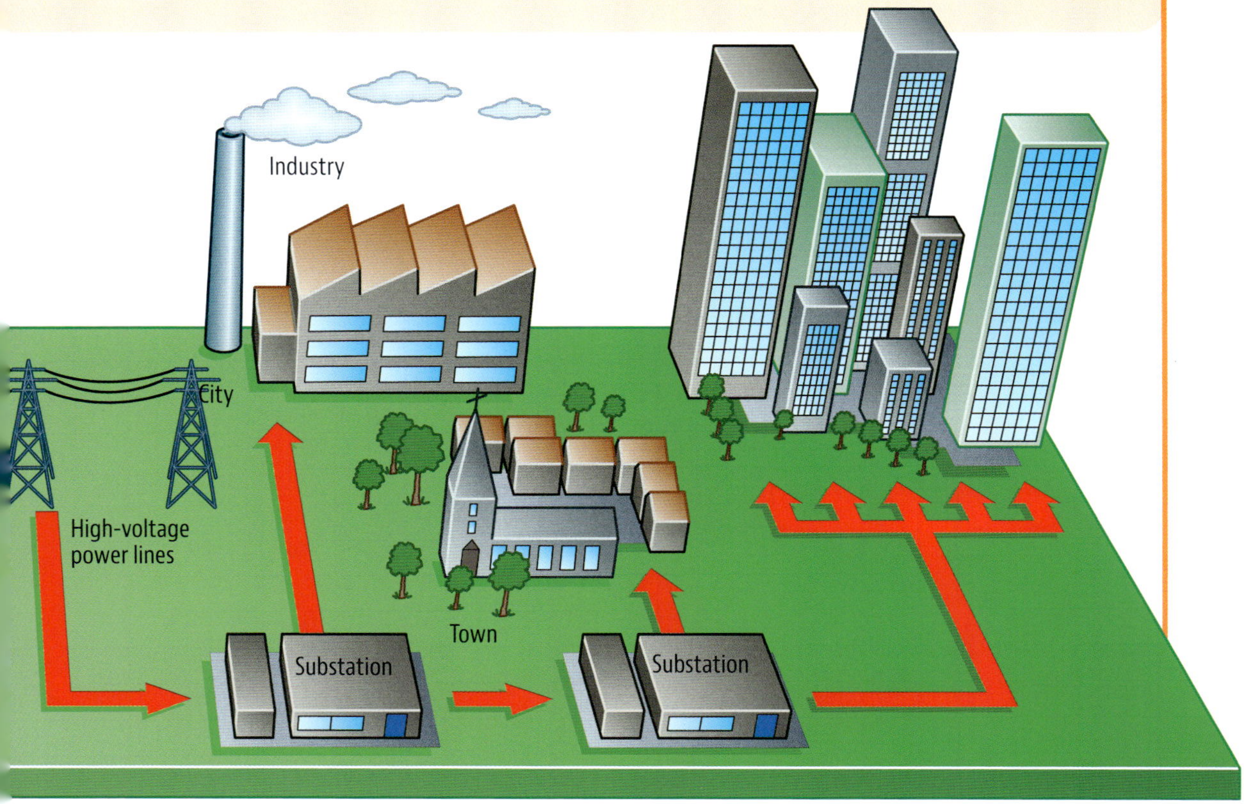

PRINCIPLE OF AC AND DC GENERATORS

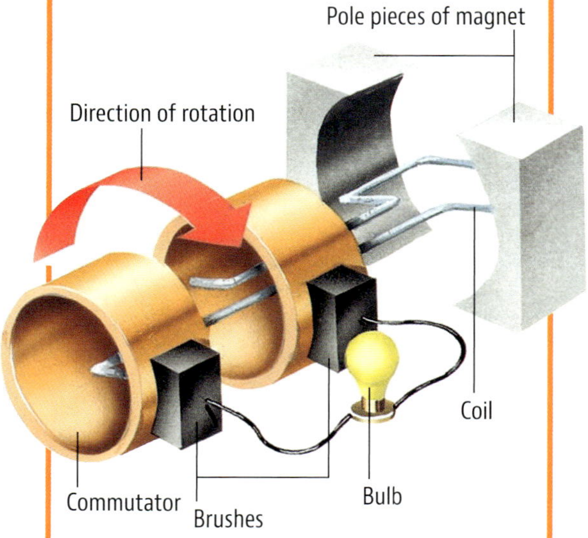

In an AC generator each side of the coil is always in contact with the same terminal so that an AC current is delivered.

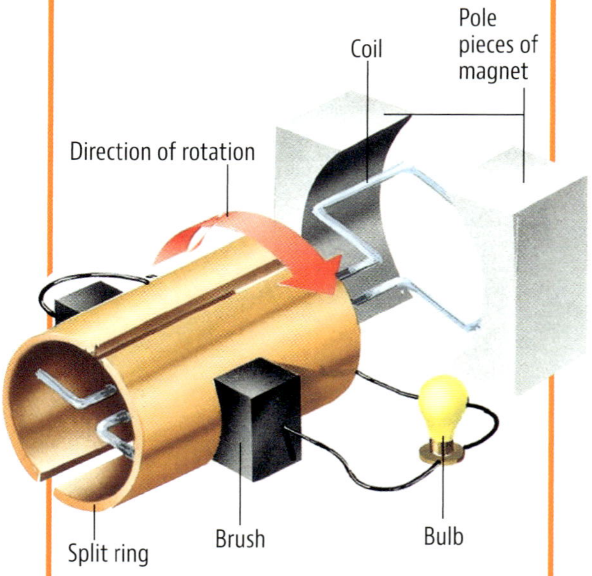

In this simplified DC generator, split-ring commutators convert the AC voltage from the rotating coil into a DC current.

A pressurized water reactor (PWR) at the Borssele nuclear power plant in the Netherlands. PWRs work on the fission principle.

Simple Generators

When a coil of wire rotates in a magnetic field the current in it is automatically AC. Think about what happens on one side of a single coil: the current in it alternately flows along it one way and then the opposite way because that side of the coil moves through the magnetic field one way in one half of its revolution and the other way in the other half. AC current can then be drawn from the coil by having a terminal of the external circuit that is permanently connected to each side of the coil. Sometimes it is advantageous to produce DC current. In this case, the connections to the generator have to be more complicated. The two terminals of the rotating coil can be in the form of the two halves of a split ring. They make contact alternately with one terminal of the external circuit and then the other. This process reverses the connection at the moment the voltage in the rotating coil

reverses. The current in the external circuit is DC, although it changes in strength during each rotation of the coil.

Complex Generators

Practical generators have to be much more complicated. These have a set of rotating coils, and the current is drawn only from the coil that is experiencing a peak voltage at that particular moment. The magnetic field is provided by an electromagnet—that is, it is generated by electric current flowing in special coils.

Nuclear Power

Electricity can also be generated from the energy released by nuclear fission or fusion. There are various designs of nuclear reactor,

GENERATING ELECTRICITY FROM NUCLEAR ENERGY

Just as in coal- or oil-fired power stations, nuclear power stations use the heat they generate to produce steam, which drives turbines. The turbines drive electrical generators.

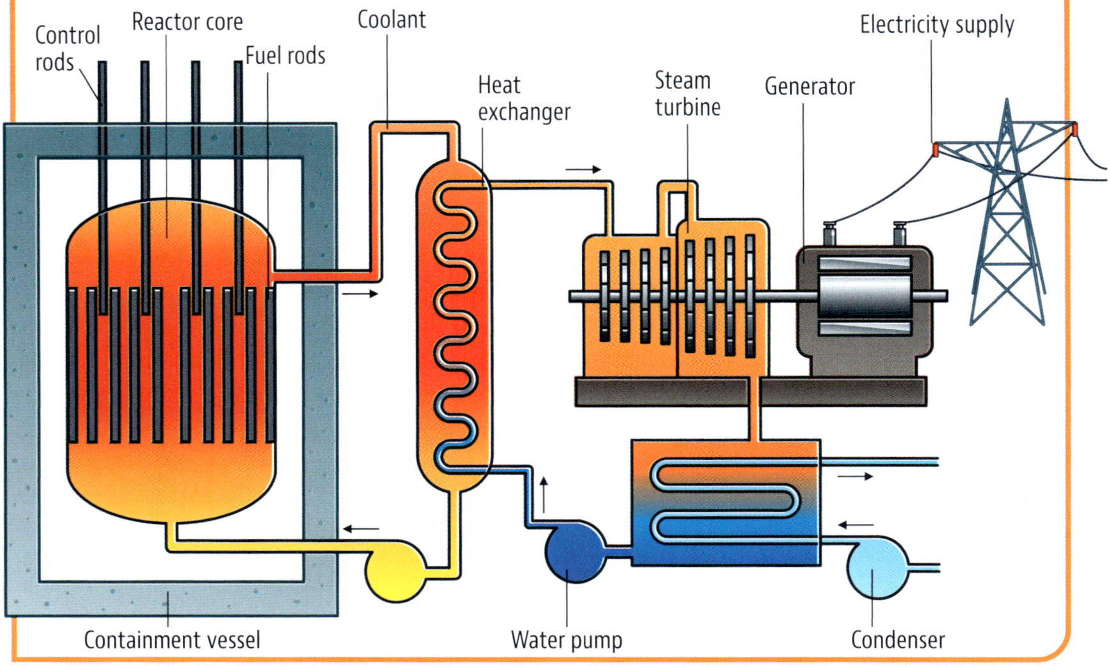

which differ mainly in the coolant they use. One of the greatest problems facing the designers of practical nuclear power plants is the radioactivity of the reactor. Neutrons from the fission processes penetrate thick layers of most materials. To shield the outside world from these neutrons, there is always a protective shield, called a biological shield, surrounding the reactor. But the coolant—the fluid that flows through the reactor and takes away its heat to be used—becomes intensely radioactive.

Reactors

In boiling-water reactors (BWRs), a type of pressurized water reactor, ordinary water is used as the coolant. It boils in the reactor, and the steam drives turbines directly. In other types of reactor, the coolant flows in a closed circuit and has no direct contact with the outside world. It is led around other pipes carrying water, which is heated to boiling point. The steam from the boiling water drives turbines. As the hot steam flows through the turbine, it makes the turbine wheels spin, driving an electricity generator.

SCIENCE WORDS

Fission: The splitting of a heavy atomic nucleus into two roughly equal parts, with the release of energy.
Fusion: The joining of two light atomic nuclei where the nuclei fuse together to create one larger, heavy nucleus and in the process release energy.
Neutron: An elementary particle that is electrically neutral (uncharged) and has approximately the same mass as the proton. Outside the nucleus it decays into a proton, an electron, and an antineutrino (the antiparticle of a neutrino).
Reactor: A device for generating energy, either by nuclear fission or by nuclear fusion.

Wind turbines harness the movement energy of wind and turn it into electricity. The wind spins the blades of a huge turbine, which spins a generator. Wind farms can be built on land or offshore.

After passing through the turbines, the steam is still hot and at quite a high pressure. It then needs to be condensed—turned back into water—to go around the circuit and be boiled again. The steam is cooled by being passed over other pipes carrying cold water from some external source, such as a nearby river or lake.

Climate Changes Everything

Today, climate change—with increasing levels of greenhouse gas (GHG) emissions causing global warming, with shifting weather patterns of floods and droughts, and rising sea levels—is having an impact on the Earth on an unprecedented in scale. Replacing the burning of finite fossil fuels with renewable energy sources is one of the ways that climate change might be slowed down. Sunlight, wind, water, and geothermal energy are clean and unlimited fuel alternatives.

Today, wind turbines are a common sight across Europe and North America.

This solar power plant in Spain has an output of 50 mW (megawatts), enough to supply electricity to around 25,000 homes. It uses panels of photovoltaic cells to generate current.

The electricity is generated in the wind turbine itself. In the United States, where wind-generated power has increased over 60 times in the 25 years between 2000 and 2024, 74,695 wind turbines operate across 41 states, Guam, and Puerto Rico. Wind power is now the largest source of renewable generating capacity in the United States.

Offshore wind farms generate more electricity than land-based turbines. This is because the offshore wind speeds are faster. A turbine in a 15 mph (24 km/h) offshore wind will generate twice as much energy as a turbine in a 12 mph (19 km/h) wind onshore. The electricity produced provides renewable energy without harming the environment.

Solar energy can also be turned into clean electricity. A large field of mirrors can be used to reflect sunlight onto a boiler, producing steam that drives a turbine in the ordinary way. Large mirror installations in the Mojave Desert use thousands of mirrors to focus sunlight onto a vessel containing a molten salt. The salt's stored heat is used to boil water and make high-pressure steam. Solar panels are another way of converting sunlight into electricity. Solar panels are coated with a photoelectric material, which directly generates an electrical current when the Sun shines on them. Photoelectric (or photovoltaic) cells are usually made from materials such as silicon. Photoelectric cells are used in solar-powered devices such as calculators. On a larger scale, they are fixed to the roofs of houses. In the U.S. and the UK, large solar farms generate clean, green electricity on a huge scale. This electricity feeds into the national grid.

In some U.S. states, electricity supplies are deregulated. Consumers choose an energy company, and that company accordingly generates more electricity. The difference to the consumer is in the prices that the companies charge, and also in the methods of generation they use. Some companies supply "green" electricity from "renewable" sources such as hydroelectricity, wind power, or solar energy. Consumers can express their preference for such sources by buying their electricity from those companies.

ELECTRIC VEHICLES

Motor vehicles powered by the internal combustion engine are the prime cause of congestion, noise, and pollution in most modern cities. Today, vehicles that are designed to run on the clean, silent power of electricity are revolutionizing the way people think about their environment.

About 95 percent of cars are still powered by gasoline or diesel. However, with global warming and environmental pollution on the increase, engineers with the major car manufacturing companies have developed vehicles that do not need to burn fossil fuels just to transport people and freight. Practical, affordable electric cars that run on battery power are available on the market for private and commercial use. Early electric cars had batteries that were very large, and which ran out of electricity quickly. Today, batteries that are more compact and capable of storing more energy have been developed, and electric cars now rival gasoline-powered cars in terms of performance. Recharging points can be found in most towns and cities, and many of the many modern vehicles can travel distances of up to 350 miles (600 km)

SCIENCE WORDS

Battery: Two or more electric cells that generate or store electricity.
Fuel cell: A type of electric cell (battery) in which a fuel (such as hydrogen) is converted directly into electricity.
Maglev: Abbreviation for "magnetic levitation," the technology for keeping vehicles such as trains a few inches above the track on which they run by means of mutually repelling magnetic fields, which overcome friction.
Superconductivity: The property of conducting electricity with no resistance at all. Some metals do this when cooled to a temperature close to absolute zero (–459.67°F / –273.15°C). New complex substances have been developed that superconduct at ever higher temperatures (although not yet as high as 32°F / 0°C).

To try and ease traffic congestion, many cities around the world invested in light-rail transit systems, with trains powered by electricity from overhead lines. This is the METRO light-rail system, which began operating in Phoenix, Arizona in 2008.

before needing to be recharged. Energy can also be stored in "ultracapacitors," which are devices that store large quantities of electricity. They are teamed with batteries in so-called hybrid systems. Some hybrid systems have energy-storage devices combined with gasoline engines. These engines are cleaner and far more economical than regular internal combustion engines.

Regenerative Braking

A feature of some electrical propulsion systems is that part of the energy normally wasted in braking can be stored and reused. Instead of conventional brakes, magnets attached to a vehicle's wheels induce electric currents in a circuit and charge up a battery, or store the energy some other way, while at the same time the vehicle is slowed. Later the stored energy can be used to boost the vehicle's speed.

Electric Streetcars

There are many vehicles in the world today that use electrical energy, not from self-contained sources on the vehicle but from the public supply, via rails and overhead lines. Streetcars supplied from overhead electrical lines were popular in many cities in the early 20th century and today are enjoying a revival. They do not have the freedom of movement of gasoline-fueled buses, but are cheaper to run and less polluting.

Electric Trains

A major use for electrical vehicle propulsion today is for railroads. Although the first subway trains were steam-powered in the 19th century, electricity was adopted as soon as it became available. This was because steam and smoke were unpleasant and even dangerous in the confines of subway tunnels. Electricity is nearly always used for the increasing numbers of light railways or rapid-transit systems being built in cities. Electricity is also the preferred system for powering long-distance trains all over Europe. However, even diesel trains are usually better described as diesel-electric, because most use their diesel engines to generate the electricity that turns the wheels.

Since 1981, France has had a network of high-speed rail services operated by electric trains called TGVs (an abbreviation of *Train à Grande Vitesse*). Each train is propelled by two power cars, one at each end. One of them collects power from overhead lines by an assembly called a pantograph, different ones for lower- and higher-voltage supplies. Part of the power is sent along cables to the other power car. Each motor delivers 1,100 kilowatts. The TGV V150 holds the record for the highest speed on any national rail system, operating at 357 mph (575 km/h).

Maglev Trains

Still faster speeds will become a reality if trains are separated from the track altogether. That can be done by setting up powerful magnetic fields in the body of the train and in the track that repel each other, a process called magnetic levitation, or maglev. Maglev trains can be supported by repulsion, with the whole train and its magnets above the track, or by attraction, with "wings" mounted on the bottom of the train to carry its magnets. They curl around under the track and are lifted up toward it. Germany was the world leader in maglev technology, developing the Transrapid system, which it has exported to China.

A maglev train on its elevated track. The first commercial maglev entered service in China in 2002, linking the city of Shanghai with Pudong International Airport.

Maglev trains are economical because there is almost no resistance between the train and the track. Also, if they use superconducting magnets, in which the magnet has been cooled to such a temperature that it has no electrical resistance, then very little energy is lost in maintaining the field. Superconducting maglev trains could become the most advanced way of harnessing magnetism to meet the world's future transport needs.

Solar-Powered Vehicles

Scientists and engineers everywhere work hard to exploit the potential of photoelectric cells. These cells are already indispensable in spacecraft, which can be powered for many years by sunlight. In space, sunlight is uninterrupted and undimmed by the Earth's hazy or cloudy atmosphere. Experimental vehicles have been developed that run on solar energy alone. Many of them compete in the biennial World Solar Challenge race from north to south across Australia. Winners have reached speeds of 55 mph (88 km/h), but this was under exceptionally favorable conditions and with machines swathed in solar cells and highly streamlined.

MAGLEV PROPULSION

Powerful superconducting magnets in the body of the train keep it suspended and centered over the track by repulsion from the traveling magnetic poles in the trackbed and at the sides of the track.

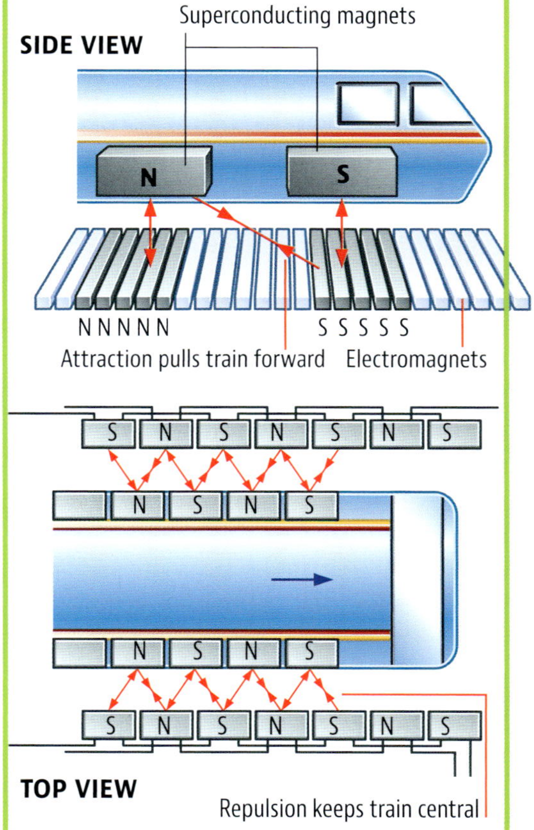

Another great innovation in motive power was the fuel cell, which is an alternative to the battery. The fuel cell generates electricity from chemical reactions, as do batteries, but the chemicals are supplied continuously from an external source. Fuel cell use a variety of fuels, including hydrogen, methane, and carbon monoxide, that react with air. Often the only waste product is water.

Developing Electric Cars

The idea of electrically-powered cars was not new. Inventors in Europe and the United States began to develop the concept of battery-powered vehicles in the early 1800s. The first successful electric car appeared in the United States in 1890. It was capable of 14 mph (22.5 km/h). In around 1898, Ferdinand Porsche invented an electric car that was powered by gas engine and electricity. However, by the 1920s, improvements in the internal combustion engine and the availability of cheap gas led to the abandonment of electric vehicles in the United States by the 1930s.

Gas shortages in the United States in the 1970s led to renewed interest in electric vehicles. However, limited top speeds of 45 mph (72.5 km/h) and a range of 40 miles (64 km) significantly limited their public and commercial appeal.

Environmental Change

Electric cars began to be taken seriously in the first part of the 21st century, with growing concerns about the effect gasoline emissions were having on the environment. In the late 1990s, the world's first hybrid electric vehicle was released in Japan. Today, plug-in (PEV), hybrid electric (HEV), and plug-in-hybrid (PHEV) vehicles are widely available. A PEV is any motor vehicle that can be recharged by plugging it into an external source of electricity. The electricity is stored in

BATTERY-ELECTRIC VEHICLE (BEV)

Electric cars store energy in a battery. Electrical energy drives an electric motor that transforms electrical energy into mechanical energy, which turns the wheels. All cars have brakes. Electric cars have regenerative brakes, which when depressed turn the energy back into electricity for storage in the vehicle's battery.

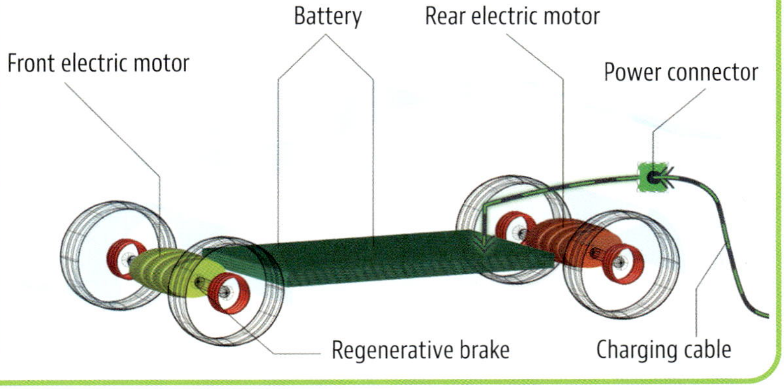

Curbside battery recharging stations are familiar sights on city streets throughout the world. The electricity often comes from renewable resources such as solar and wind power.

SCIENCE WORDS

n-type semiconductor:
A semiconductor material in which current consists mostly of electrons in motion.

p-type semiconductor:
A semiconductor material in which current consists mostly of moving holes.

Photoelectric cell (photocell):
A device that uses light or other electromagnetic radiation to produce a current or voltage. Photocells are commonly used in automated restrooms and burglar alarms.

rechargeable batter packs. An HEV is a vehicle that combines a conventional power source (such as fossil fuels) with some form of electric propulsion. And a PHEV is a vehicle in which the charge to the batteries is provided through both a plug-in source and electric propulsion. Hybrid vehicles were a halfway house between conventional gas cars and the fully electric cars (BEVs) that some people drive today. Solar-powered fast-charging "filling" stations have cut down the time that is required to recharge batteries on the go.

SOLAR POWER

This solar-powered vehicle (seen from above), covered in photoelectric cells, traveled in the bright sunshine of the Australian deserts at an average speed of 30 mph (50 km/h). It took part in the World Solar Challenge race from Darwin to Adelaide.

Solar-powered vehicle

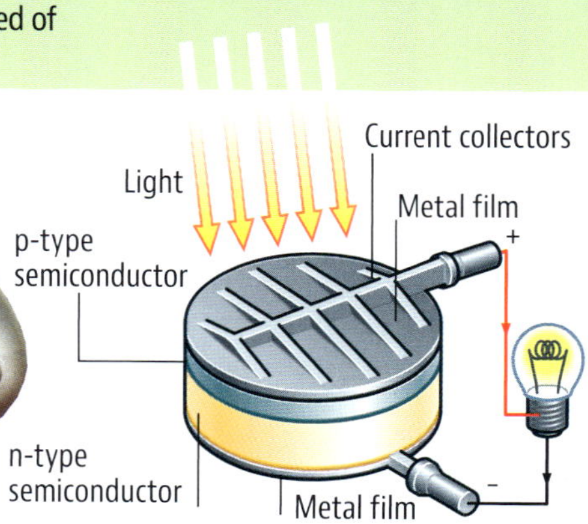

Each solar cell consists of layers of two types of semiconductor, called p-type and n-type, sandwiched between metal films.

BATTERIES

Electric and hybrid vehicles are powered by batteries and fuel cells. The historical development of batteries and different types of fuel cell began in around 1800, and made today's clean, renewable energy sources possible.

Batteries are a way of converting chemical energy into electrical energy. There are two main kinds of battery, disposable (primary cell) and rechargeable (secondary cell). In a primary cell, the chemicals it contains are gradually used up. When they have been consumed, the battery no longer produces electric current.

A secondary cell is similar except that when the chemicals have been used up, the battery can be recharged. By pushing an electric current through the cell (as in electrolysis), the chemicals are regenerated, and the battery will work again. Because a secondary cell effectively stores or accumulates the charge put into it, this type of battery is also called an accumulator. An accumulator can be discharged and recharged over and over again.

In around 1800, the Italian physicist Alessandro Volta (1745–1827) made the first battery. It consisted of alternating disks of copper and zinc, separated by disks of cloth, leather, or paper soaked in salt solution. A length of wire was attached to the topmost copper disk, and another wire was connected to the bottom zinc disk. Because it consisted of a stack of disks, the battery came to be called the "voltaic pile." The pile works by electrochemical processes. The copper metal loses electrons to the solution between the disks and forms copper ions. At the same time, zinc dissolves in the solution and releases hydrogen gas. If the wires are joined in an electric circuit, current flows from the positive copper disk through the circuit and into the negative zinc disk.

SCIENCE WORDS

Anode: An electrode with a negative charge, at which electrons leave a voltaic cell.
Cathode: An electrode with a positive charge, at which electrons enter a voltaic cell.
Electrolysis: The decomposition of an electrolyte by electric current, using two electrodes.
Polarization: The process that causes bubbles of gas to appear on one of the electrodes of an electric cell, stopping the cell from working.

Alessandro Volta became internationally famous after he invented the voltaic pile in about 1800. Here, Volta (on the left) demonstrates his battery to the French Emperor Napoleon.

The next stage in the development of the battery was a simple cell made by immersing a copper plate (cathode) and a zinc plate (anode) in a solution of dilute sulfuric acid. When a length of wire connects the plates, current flows from the copper to the zinc. Negative sulfate ions from the acid electrolyte travel to the zinc electrode, where the zinc dissolves to form zinc ions. At the same time, positive hydrogen ions travel to the copper electrode, where they are discharged to produce bubbles of hydrogen

ALESSANDRO VOLTA

Volta's full name was Count Alessandro Giuseppe Antonio Anastasio Volta. He was born in 1745 in Como, Italy. Unlike his male relatives, who nearly all became priests, Volta decided to study electricity. In 1774, he invented the electrophorus, a device that could produce and store a static electric charge. His best known invention was the first battery to produce a steady electric current. This consisted of a pile of alternating disks of copper and zinc separated by pieces of cloth soaked in salt solution. It became known as the voltaic pile. Volta died in 1827, some years before the unit of potential difference was named the "volt" in honor of his achievement.

FIRST BATTERY

The first battery consisted of a stack of disks and so was called a voltaic pile after Alessandro Volta, the Italian scientist who invented it. The pile had alternate disks of copper and zinc separated by disks of cloth soaked in salt solution.

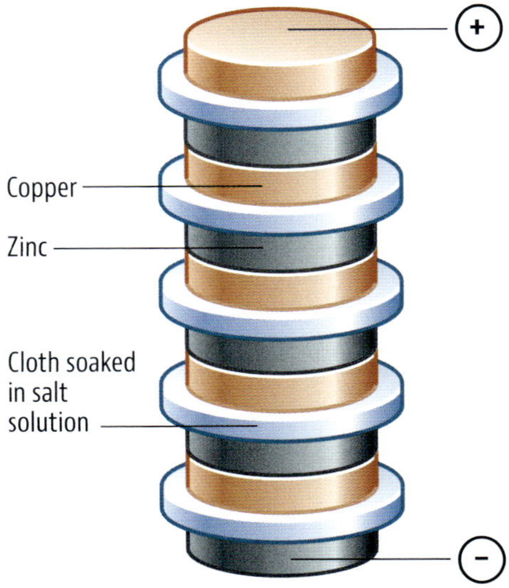

gas. As soon as hydrogen bubbles cover the copper electrode, the action of the cells slows down and eventually stops. When this happens, the cell is said to be polarized.

Practical Batteries

The first attempt to find a way around the problem of polarization was made by the British physicist John Daniell (1790–1845). In the so-called Daniell cell, the copper electrode (cathode) takes the form of a cylindrical copper container. The zinc electrode (anode) is a metal rod inside a porous earthenware pot containing dilute sulfuric acid. The porous pot stands inside the copper container, and between the two is a solution of copper sulfate.

When a length of wire connects the electrodes, hydrogen ions travel toward the inside of the copper container. Because of the presence of the copper sulfate solution, instead of hydrogen gas being released, copper metal is deposited on the container. There is no polarization, and in this cell the copper sulfate acts as a depolarizer. At the zinc electrode, zinc sulfate is again formed as in the simple cell. A Daniell cell produces about 1.1 volts.

Leclanché Cell

A different depolarizer is used in a cell invented in 1865 by the French scientist Georges Leclanché (1839–1882). In the Leclanché cell, a zinc anode dips into a container of ammonium chloride solution. The cathode is a carbon rod surrounded by a depolarizer of manganese dioxide. Powdered carbon is mixed with the manganese dioxide to make it a better conductor. Hydrogen is still produced at the cathode, but the manganese

DRY BATTERY

A dry battery as used in a flashlight is a type of Leclanché cell that has a paste of ammonium chloride as the electrolyte.

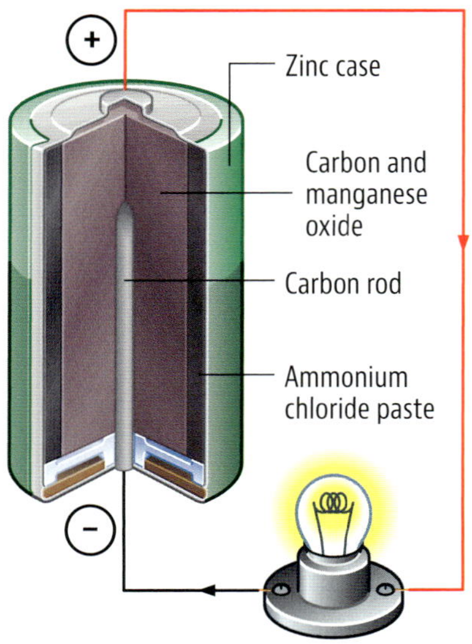

Lead-acid car batteries for scrap. Almost all the lead can be reclaimed. To encourage recycling, many U.S. states charge a deposit on new batteries, which is refunded when the exhausted battery is returned.

THE DANIELL CELL

The Daniell cell is a type of voltaic invented by John Daniell (1790–1845). The anode is zinc metal and the cathode is solid copper. The cell has two electrolytes. The cathode is in copper sulfate solution, while the anode is in sulfuric acid. The two liquids are separated by a porous barrier. The zinc atoms in the anode are oxidized. They lose two electrons and form zinc ions (Zn^{2+}). These ions dissolve in the acid. The copper ions (Cu^{2+}) in the copper sulfate are reduced. They pick up electrons from the cathode and form copper atoms, which join on to the cathode. The electrons released in the anode move through a wire to the cathode, producing an electric current of 1 volt.

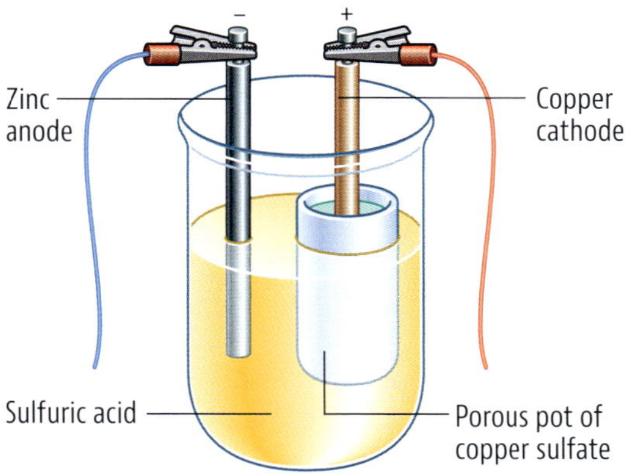

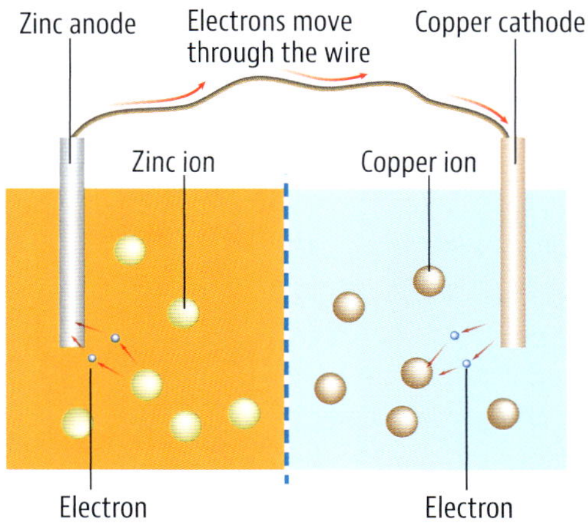

dioxide gets rid of it by converting it to water. A Leclanché cell produces about 1.5 volts. To get a higher voltage, several cells are joined in a group, or battery—which is why primary cells are called batteries.

Away with Water

The cells described so far are all wet cells—they have various solutions as their electrolytes. This does not make the batteries very portable. The problem is overcome in a dry battery, which contains no liquid. It is a type of Leclanché cell, but the electrolyte is a paste of ammonium chloride and gum. The zinc anode forms the case of the battery, and manganese dioxide and carbon again surround the central carbon cathode. Dry

cells of this kind are commonly used in flashlights and portable radios. There are many other types of dry battery, including the mercury oxide cell, which is used in many electronic watches. It produces 1.35 volts. A Leclanché-type cell with zinc chloride instead of ammonium chloride works better at lower temperatures. A nickel–cadmium cell is different because it can be recharged after it has run down by connecting it to an external supply of current. It is therefore a type of secondary cell, or accumulator.

Accumulators

Gasoline-powered cars use large lead-acid batteries, sometimes called accumulators, to provide electricity. The usual kind has anodes made of lead and cathodes made of lead covered with a layer of lead oxide. The electrolyte is sulfuric acid. When the accumulator is in use, sulfate ions react with the lead anode to produce lead sulfate and release electrons. At the cathode, hydrogen ions (from the acid) and sulfate ions react with the lead oxide to produce lead sulfate and water. This reaction needs a supply of electrons. To recharge the accumulator, current from an outside source is passed through the battery in the opposite direction. This reverses the reactions at the electrodes, re-forming lead and lead oxide. The accumulator is then ready for use again.

In a conventional automobile, the battery is used mainly to start the gas engine. Once the motor is running, an alternator supplies current to recharge the battery. A lead–acid cell produces about 2 volts. A car battery has six cells, giving a total voltage of 12 volts.

There are several other kinds of accumulator. The nickel–iron accumulator

HOW A GASOLINE CAR BATTERY WORKS

A common accumulator has electrodes of lead and lead oxide in an electrolyte of sulfuric acid. It produces about 2 volts. A car battery has six lead–acid cells, giving a total voltage of 12 volts. When the battery has run down, connecting it to an outside source of current recharges it.

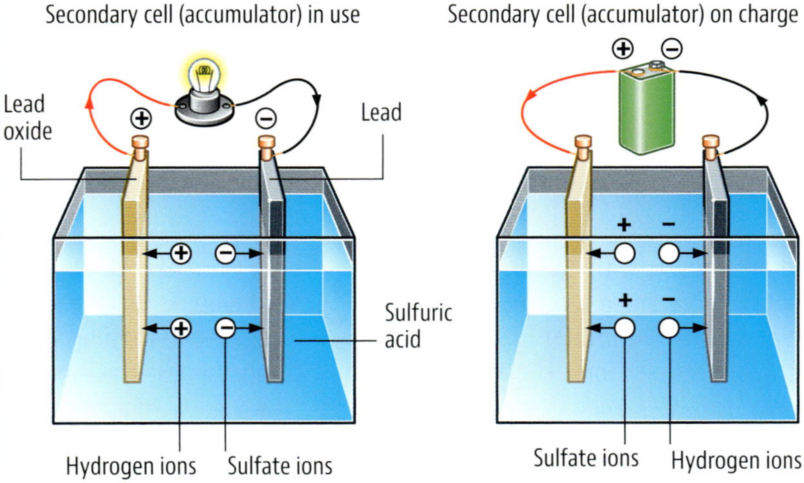

Fuel cells in some cars have platinum catalysts, which produce energy from hydrogen and oxygen.

was invented by Thomas Alva Edison (1847–1931) and known also as a NIFE cell (after the chemical symbols for nickel, Ni, and iron, Fe). The electrolyte is potassium hydroxide.

Fuel Cells

A fuel cell is a type of battery that converts a fuel directly into electricity. One type of fuel cell uses hydrogen gas as fuel. It also requires a supply of oxygen. The electrolyte is potassium hydroxide solution. Hydrogen reacts with hydroxyl ions to produce water and a supply of electrons. Oxygen reacts with water to produce a new supply of hydroxyl ions. The overall reaction is the chemical combination of hydrogen and oxygen to form water and generate electricity.

A hydrogen fuel cell can power electric cars in place of internal combustion engines. These fuel cell cars emit only water vapor in place of toxic exhaust. The next generation of electric vehicles might create their own electricity from hydrogen—resulting in zero greenhouse-gas emissions. Many scientists believe hydrogen is the fuel of the future. However, the real choice for the foreseeable future is between gasoline cars and battery-electric vehicles (BEVs).

LITHIUM-ION BATTERIES

You can quickly charge your phone, laptop, or electric car, and use it for many hours, thanks to the lithium-ion (Li-ion) batteries inside. They have many advantages over older types of rechargeable batteries: they are lighter, less expensive, store more energy, and don't use harmful chemicals such as cadmium or lead. They have a positive terminal (made from a lithium compound), a negative terminal (made from graphite), and an electrolyte in between. When the batteries are charging, the lithium compound splits into ions, which move toward the negative electrode and stay there. When the ions have all moved over, the battery is fully charged. When the battery discharges, the lithium ions move back again. This process can be repeated over and over; a lithium-ion battery can be recharged many hundreds of times.

SCIENCE WORDS

Electrode: A metal plate or carbon rod that carries electrons into or out of a battery (cell), electrolysis cell, or vacuum tube.
Electrolyte: A liquid that conducts electricity, as in a battery or in electrolysis.
Primary cell: A type of electric cell in which chemical reactions produce electricity. It cannot be recharged.
Secondary cell: A type of electric cell in which chemical reactions produce electricity. It can be recharged.

GLOSSARY

AC motor An electric motor that is operated by alternating current.

Alternating current (AC) Electric current that flows first in one direction, then in the other, alternating many times each second.

Anode An electrode with a negative charge, at which electrons leave a voltaic cell.

Armature The current-carrying wire coils in a motor or magnetic device that produce force.

Bar magnet A bar-shaped permanent magnet, with a magnetic pole at each end.

Cathode An electrode with a positive charge, at which electrons enter a voltaic cell.

Coil A spiral of wire through which an electric current flows.

Current A flow of electric charge (electrons). Current may be supplied by the domestic electricity supply, generators, or batteries.

DC motor An electric motor that is operated by direct current.

Direct current (DC) Electric current that flows in one direction all the time, though it may vary in strength.

Domain A small region in a magnetic material in which the magnetic fields of individual atoms all point in the same direction, making the domain into a single small magnet.

Electrode A metal plate or carbon rod that carries electrons into or out of a battery (cell), electrolysis cell, or vacuum tube.

Electrolysis The decomposition of an electrolyte by electric current, using two electrodes.

Electrolyte A liquid that conducts electricity, as in a battery or in electrolysis.

Electromagnet A device that develops a magnetic field when electric current is passed through it. It consists of a coil with a core.

Electromagnetism The interlinked phenomena of electricity and magnetism. Every electric current generates a magnetic field, while changes in a magnetic field can cause a current to flow.

Field The pattern of magnetic, electric, gravitational, or other influence around an object.

Fission The splitting of a heavy atomic nucleus into two roughly equal parts, with the release of energy.

Fuel cell A type of electric cell (battery) in which a fuel is converted directly into electricity.

Fusion The joining of two light atomic nuclei where the nuclei fuse together to create one larger, heavy nucleus and in the process release energy.

Induction motor An electric motor in which a moving magnetic field induces (generates) current in another component, which moves to follow the field. The moving part may revolve or move in a straight line.

Insulator A material that is a poor conductor of electric current.

Line of force An imaginary line where its direction at any point in a field shows the direction of the field at that point.

Lithium-ion battery A rechargeable battery that uses lithium ions as the primary component of its electrolyte.

Lodestone A naturally magnetic iron ore, formerly used to make magnetic compasses.

Magnetic compass A device in which a freely rotating magnetized needle indicates the direction of magnetic north.

Magnetic lens An arrangement of electromagnets that can focus a beam of electrons or other charged particles, as in an electron microscope.

Magnetic north The direction in which a magnetic compass points (in the northern hemisphere).

Magnetic resonance imaging (MRI) The technique of making pictures of the inside of something, for example, the living human body.

Magnetization The process of making an object or material into a magnet.

Magnetosphere The region around the Earth or other celestial body, such as Jupiter, in which its magnetic field is stronger than the field in surrounding space.

Motor A machine that converts energy, usually electrical energy, into motion.

Neutron An elementary particle that is electrically neutral (uncharged) and has approximately the same mass as the proton. Outside the nucleus it decays into a proton, an electron, and an antineutrino.

Photoelectric cell (photocell) A device that uses light or other electromagnetic radiation to produce a current or voltage.

Polarization The process that causes bubbles of gas to appear on one of the electrodes of an electric cell, stopping the cell from working.

Pole One of the two regions of a magnet where the strength of the field is greatest. Lines of force diverge (radiate out) from one pole and converge on the other.

Primary cell A type of electric cell in which chemical reactions produce electricity. When the chemicals are used up, the cell is discharged and cannot be recharged.

Relay A device that is activated by changes in an electric current in a circuit, causing it to switch a second electric circuit on or off. Relays often incorporate electromagnets.

Rotor In electromagnetic technology, the rotating coil in a motor or generator.

Secondary cell A type of electric cell in which chemical reactions produce electricity. Unlike a primary cell, it can be recharged.

Solenoid A current-carrying coil of wire.

Stator The stationary part of a motor or generator.

Superconductivity The property of conducting electricity in which there is no resistance at all.

Torque Any force or system of forces that cause rotation.

Transformer A device that increases or decreases the voltage of alternating current.

Turbine A machine in which the energy of a moving fluid or air flow is converted into mechanical energy by causing a bladed rotor to rotate.

Van Allen belts Another name for the Earth's radiation belts.

Voltage (also called electromotive force, EMF) The potential difference in electrical charge between two points in an electrical field. It is measured in volts. The greater the voltage, the higher the force moving two charges between the points.

FURTHER RESEARCH

Books

Christopher Cooper. *The Basics of Magnetism*. NY: Rosen Young Adult, 2024.

Al-Khalili, Jim. *The Physics Book* (DK Big Ideas). New York, NY. Dorling Kindersley, 2023.

Holzner, Steve. *Physics for Dummies*. Hoboken, NJ. Wiley, 2021.

Web Sites

explainthatstuff
https://www.explainthatstuff.com/magnetism.html
Magnetism explained, with numerous links, videos, and further resources.

How Stuff Works
https://electronics.howstuffworks.com/motor.htm
How electric motors work.

National Geographic
https://education.nationalgeographic.org/resource/magnetism/
Article about Magnetism, plus links to information on Magnetic north.

INDEX